Issued under the authority of the
(Fire and Emergency Planning [

CW00621778

Fire Service

Volume 2
Operations

Aircraft Incidents

0 3 DEC 2001

0 3 DEC 2001

HM Fire Service Inspectorate Publications Section

London: The Stationery Office

ISBN 0 11 341192 8

Cover photographs:
    British Airports Authority Fire Service and West Midlands Fire Brigade

Printed in the United Kingdom for The Stationery Office
J99439 12/99 C50 5673

Aircraft Incidents

Preface

Firefighters have to effect rescue, fight fires and carry out special services at a wide range of transport incidents. This book looks specifically at incidents involving aircraft.

Serious aircraft accidents are relatively rare but when they do occur they can cause severe problems for firefighters. The degree of severity of any aircraft accident will depend to a large extent on the location of the incident; e.g., in a town, in remote countryside or at an airport; the size of the aircraft, the number of passengers on board and the nature of its cargo.

There will always be a serious risk of an outbreak of fire. The uncontrolled crash of any aircraft will inevitably result in its passengers, crew and any others in the immediate vicinity receiving serious crushing injuries, fractures, penetrating and other impact trauma injuries, whether or not a fire breaks out. Firefighters will be faced with difficulties of gaining access and the dangers posed from the aircraft's components, its contents and the materials used in its construction.

Such incidents generate intense media attention where the operations of the emergency services are rigorously scrutinised. Whilst much of this concentration is approving it will invariably focus on the preparedness of the emergency services and their operational effectiveness.

This book examines the general features of aircraft and airports and considers operational principles, liaison and planning for incidents.

This book replaces:

The Manual of Firemanship, Book 4, Part 1; Incidents Involving Aircraft and Part 6B, Chapter 4.

Dear Chief Officer letter 7/1993

Dear Chief Officer letter 10/1991 Item 6

Dear Chief Officer letter 6/1992 Item 4

Technical Bulletin 3/1954

Aircraft Incidents

Contents

The Airbus A3XX is due to enter service in 2004 with a passenger capacity of up to 960. This picture with three double-decker buses gives a clear indication of the size involved and potential problems for firefighters.

(Photo: British Aerospace)

Aircraft Incidents

Introduction

The preface of this book has already pointed out the serious problems which aircraft incidents can cause. This situation is unlikely to improve as aircraft construction develops and the number of aircraft of different types, sizes and usage increases.

Aircraft accidents remain relatively rare, which for the Fire Service means that it is difficult to acquire experience in this specialist field. Nevertheless, accidents will occasionally occur, and can happpen anywhere, so any firefighter may have to cope with such an incident.

To do so successfully, a firefighter will need a basic knowledge of the following:

● Aircraft Construction and Design
● Aircraft Type(s)
● Whether Civilian or Military.

In each category there are many variations and therefore it would be impractical, if not impossible, to describe in detail each individual type and variation that the firefighter may encounter.

This book describes the basic principles of aircraft construction and design and the main features that are commonly found, and it gives some examples. It discusses fixed firefighting and escape provisions and the special features of military aircraft. It further considers operational principles, liaison, planning and safety.

Aircraft can crash anywhere; the severity and difficulty of the incident will depend to a considerable degree on the location.

Accidents off the airport give rise to problems such as:

● Notification
● Locating the aircraft
● Mobilisation of resources (personnel and equipment)
● Access
● Media attention

These are just some examples of the problems that may arise. The accumulation of personnel and equipment, particularly in bad weather over difficult terrain, can be extremely involved and time consuming.

Accidents on airports may not be as severe or difficult since there should be a prompt attendance and rapid build up of personnel and equipment. Accidents are, however, more likely to occur on or near airports.

It is important, therefore, that Brigades ensure that their firefighters are familiar with airports, their layouts, facilities, equipment, organisations, operations and the Airport Fire Service and its resources.

This book gives a brief general description of these and of the controlling authorities. Firefighters must supplement this information by an active and robust liaison programme between themselves and the Airport Authorities within their own areas.

It must be stressed that wherever an accident occurs it will present the Fire Service with special problems. Pre-planning and liaison will be especially important, particularly where an accident amounts to a major disaster.

The remaining concerns of the book are with rescue and firefighting techniques on and off airports, and the many special considerations firefighters will have to bear in mind during such operations. These include the problems of entry; the handling of crew and numerous passengers who may be involved; the risks inherent in the components utilised in the construction of aircraft; dangerous cargo; and the special risks associated with military aircraft including weapons.

Aircraft Incidents

Aircraft Incidents

Chapter 1 – Design and Construction of Fixed-Wing Civil Aircraft

1.1 General

Aircraft vary greatly in size and design, ranging from the ultra light single and two seaters to the four engine "stretched" passenger aircraft seating 600 or more people. (See page viii.) A fuel load of 200,000 litres is not uncommon and speeds of up to 1000 kph are usual for large civil aircraft. The construction of aircraft varies with their projected use. It is not intended to describe the older type of medium size aircraft because by virtue of their design, they do not represent so great a problem to firefighters in their passenger capacities, fuel loads and access to the aircraft following an accident.

However, it must be recognised that older aircraft types during maintenance and refurbishment, will invariably be retrofitted with contemporary equipment and materials, customarily including composites. The use of these materials is featured within this chapter with the special hazards they present described in Chapter 8 – "Special hazards in Aircraft Incidents".

1.2 Body Construction

1.2.1 Fuselage

The tapering shape of the fuselage is formed by a series of vertical metal frames placed transversely from nose to tail. Metal stringers running horizontally along the length of the fuselage are spaced around the circumference of the frames. (See Figure 1.1.) Internally placed stringers made of stronger and thicker metal, which are called longerons, are continuous along the length of the fuselage and serve as attachment points for the cabin floors, cargo holds etc. The skin is not merely a sheet metal covering but is stressed according to the load it must take and contributes to the total rigidity of the airframe. The skin panels may either be riveted or bonded to a number of stringers to form a separate assembly, which is then riveted to the frames. When the skin is bonded there is no external indication of the underlying framework. Due to pressurisation, the fuselage of most modern aircraft consists of a double skin with a suitable insulating material interposed. The system of construction described is usually referred to as "Semi-Monocoque".

1.2.2 Mainplane (Wings)

Tapering metal spars run from the centre section to the wingtips, or from wingtip to wingtip running through the fuselage. Their vertical height forms

Figure 1.1 The basic design of a stressed skin construction in the fuselage of an aircraft.
(Photo: British Aerospace)

the thickness of the wings. Some aircraft have only two spars but many have several which vary in length according to the wing configuration, i.e., swept, delta, crescent etc. Short metal struts known as ribs are closely spaced at right angles to the spars and form the profile of the aerofoil of the wings. The whole construction is covered by the skin, which again, forms part of the structure and is either bonded or riveted to the spars. (See Figure 1.2.)

1.2.3   Tail unit (empennage)

The tail unit is constructed in much the same way as the mainplane. It consists of the tailplane which aids longitudinal stability and the tailfin which aids directional stability. Moveable control surfaces on the unit consist of elevators to control dive and climb and the rudder to control the direction of the aircraft.

Figure 1.2 Mainplane Construction

(Photo: British Aerospace)

1.2.4 Metals used in Aircraft Construction

Firefighters should try to obtain knowledge of the type of metals used in aircraft structures and their likely locations because their reaction to impact, fire and cutting, will have a pronounced effect upon rescue and firefighting.

The types of metal most commonly used in aircraft construction are:

● **Aluminium Alloys**

This is the most common of the metals used in aircraft construction. Their composition varies depending upon where they are used, i.e., skin surfaces, formers, stringers, spars, etc.

Aluminium alloys can be readily cut with axes, hacksaws or powered cutting tools. Aluminium alloys are good conductors, quickly transmitting heat and equally rapidly cooled by water spray or foam.

● **Magnesium and its Alloys**

Magnesium alloy is a light, strong metal that can be found in engine mounting brackets, crankcases in piston engines, compressor casings of turbine engines and various other areas where strengthening may be required and where bulkiness is no object.

Magnesium and its alloys once ignited, can prove difficult to extinguish as they react violently to most firefighting agents. Magnesium and its alloys are unlikely to be found in areas where forcible entry may be necessary.

● **Titanium Alloys**

Titanium alloy is used where greater strength or resistance to heat is required. Its primary use is in engine firewalls, tailpipe casings and turbine engine blades. In some high-speed aircraft titanium is used to make major components such as the leading edges of wings etc.

Titanium is difficult to ignite, but once ignited, it can prove difficult to extinguish. Class "D" dry chemical powders may be used, if available.

It is difficult to cut other than thin sections. However, sparking may result from prolonged friction with the use of powered cutting tools.

● **Stainless Steel**

Stainless steel is used where greater strength and rigidity is required such as in frames which act as attachments for the mainplane, beams supporting engines, parts of the undercarriage and for reinforcement of skin surfaces.

An indication of the engineering complexity and meticulous attention to detail demanded in the construction of a modern aircraft is shown in Figure 1.3.

● **Composite Material**

Although not a metal, the use of composite materials in aircraft construction is common. Composite materials are also known collectively as Man Made Mineral Fibres (MMMF). The term MMMF describes a wide range of materials which utilise the inherent strength and durability of woven fibres bonded together with resins. Carbon Fibre Reinforced Plastics (CFRP), Aramid Reinforced Plastic (ARP), Glass Fibre Reinforced Plastic (GFRP) and Kevlar are all common names used to describe these materials.

Figure 1.4 gives an indication of the quantity of MMMF used in modern aircraft construction and where it may be found.

The use of such materials present certain hazards and dangers. These are described more fully in Chapter 8 – Special Hazards in Aircraft Incidents.

1.3 Aircraft Engines

1.3.1 Piston Engines

Piston engines are rare on large aircraft and the hazards they present are relatively slight, although one potential danger is the presence of a propeller. Broken fuel or oil lines, especially if they are close to hot exhausts, or damaged electrical wiring is the most likely cause of fire. Provided that the fire has not penetrated the fire resistant bulkhead which separates the engine from the adjoining parts of the

Figure 1.3 Advanced engineering and meticulous attention to detail is demanded in the construction of a modern aircraft.

(Photo: British Aerospace)

aircraft, the application of Halon, CO_2, or Water-fog will usually be sufficient to extinguish any fire.

These types of engines may be divided into three main categories:

- In-line
- Flat (horizontally opposed)
- Radial

1.3.2   Turbine engines

These types of engines can be divided into three main categories:

- Turbojet
- Turbofan
- Turboprop

Each of these present their own special problems when dealing with fires within their engine com-partments, for example, the Turboprop has a drive shaft extending forward to drive a propeller.

For firefighters the main problems associated with these types of engines can be fuel spilling from broken pipes in a spray or mist form or an internal fire within the combustion chamber following a crash.

Most modern jet engined aircraft have engines located under the wing and are therefore fairly low to the ground and are readily accessible. Some air-craft have additional rear mounted engines and these present problems for access and the applica-tion of media.

Firefighters must bear in mind that owing to their weight and momentum, compressors and other rotary parts of an aircraft engine will continue to run for a considerable time after the engine has been shut down.

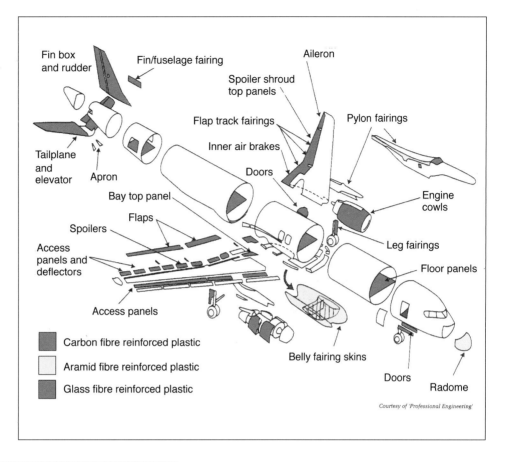

Figure 1.4
An indication of the quantity of MMMF used in modern aircraft construction and where it may be found.

Fin box and rudder
Fin/fuselage fairing
Aileron
Spoiler shroud top panels
Pylon fairings
Flap track fairings
Inner air brakes
Tailplane and elevator
Apron
Doors
Engine cowls
Bay top panel
Leg fairings
Flaps
Spoilers
Floor panels
Access panels and deflectors
Access panels
Belly fairing skins
Doors
Radome

Carbon fibre reinforced plastic

Aramid fibre reinforced plastic

Glass fibre reinforced plastic

Courtesy of 'Professional Engineering'

Danger areas will vary with contrasting aircraft, the type, size and location of engines fitted. In general, firefighters should keep clear of engine intakes and exhaust as serious or even fatal injury may result.

Piston engine and turboprop aircraft also have propellers that can inflict severe injury or kill.

The majority of fires within aircraft engines will usually be associated with the accessory section. This section contains the following:

● Fuel pumps
● Fuel lines
● Hydraulic pumps
● Hydraulic lines
● Oil pumps
● Oil lines
● Gearbox (if turboprop)
● Electrical generators

If any of the fuel pumps or fuel lines leak, the fuel or lubricants will be released in mist form under high pressure.

1.4 Fuel and Fuel Tanks

1.4.1 Types of Fuel

Fuels for aviation use broadly fall into two types, Petrol (Gasoline) and Kerosene.

(a) Petrol (Gasoline, and Avgas)

Petrol, also known as gasoline or Avgas is used in piston engine aircraft, the fuel having different grades. These grades are characteristic of the octane rating of the fuel, which means that their composition differs depending on the engine compression ratio.

The grade (octane rating) of gasoline will have no bearing on the flammability of the fuel. For further details see the "Physical Properties of Fuel".

(b) Kerosene

These fuels are used in turbine engine aircraft and fall into two categories:

(i) *Avtur (Jet A1)*

Avtur or jet A1 is the most widely and commonly used kerosene fuel. Avtur will not ignite under normal temperature and pressure but may do so if sprayed into a hot engine following a crash.

(ii) *Avtag (Jet B)*

Avtag is a wide cut fuel of approximately 60% gasoline and 40% kerosene. Avtag will readily ignite at normal temperature and pressure having the characteristics of Avgas.

Physical Properties of Fuel

Fuel	Flash point	Auto Ignition Temp.	Rate of Flame spread M/min
Avgas	-40°C	450°C	215-245 m
Avtur	38°C	245°C	30 m
Avtag	-23°C	250°C	215-245 m

Limits of Flammability

Fuel:
Avgas   1.4% – 7.6%
Avtur   0.7% – 5.8%
Avtag   0.8% – 5.0%

Avgas and Avtag will ignite at normal temperatures and pressure. Avtur will NOT ignite under these conditions. However, if the fuel is sprayed onto a hot engine or hot aircraft components, it may ignite. Once ignited it will burn as readily and produce as much heat as a fire involving any of the other two fuels.

The rate of flame spread is however greatly reduced.

It is important to remember that any fuel in MIST FORM will readily ignite and propagate the fire faster.

All aviation fuels are corrosive, irritant, toxic and can contaminate other materials.

1.4.2   Fuel Tanks

Fuel is carried in a number of structurally separate but interconnected tanks and can be found in the wings, fuselage and the tail plane of an aircraft. (See Figure 1.5.) The principal types of fuel tank are:

(a)   Rigid tanks

These are usually constructed from sheet aluminium with internal baffles. The baffles reduce surge and help to strengthen the tank. They are often covered with fabric, and have a vent pipe, an overflow, a base sump and a fuelling orifice.

These tanks are usually set in cradles within the wings or fuselage and are strapped in and bonded to the aircraft to prevent the formation of static electricity.

(b)   Integral tanks

Integral tanks use the aircraft airframe compartments for storing fuel. These are invariably found in the wings but may additionally be located in the fuselage and tailplane area. The use of the aircraft's airframe produces the lightest of fuel tanks and the most commonly used in the larger contemporary passenger carrying aircraft.

Any aircraft accident can readily distort or damage such tanks, or split joints with a consequential release of fuel.

(c)   Flexible tanks

These tanks are flexible bags made of plastic, nylon or neoprene rubber or other man made material, which are fitted into the wings or the fuselage and secured by press-studs.

These tanks have the advantage of being very resistant to shock and may not suffer damage in an accident unless they are cut on jagged metal surfaces.

However, because of their construction materials, they are flammable and give off toxic vapour when burning.

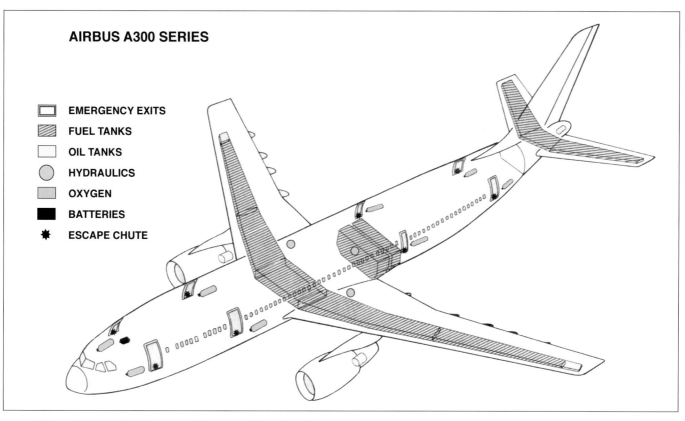

Figure 1.5 Fuel Tanks on a modern aircraft are found in wings, fuselage and tail plane.

AIRBUS A300 SERIES

- ☐ EMERGENCY EXITS
- ▨ FUEL TANKS
- ☐ OIL TANKS
- ⬤ HYDRAULICS
- ▦ OXYGEN
- ■ BATTERIES
- ✳ ESCAPE CHUTE

(d) Auxiliary tanks

Many aircraft can be fitted with extra fuel tanks and generally these will be found under the fuselage or mainplane or at the wingtips.

Wingtip fuel tanks are usually made of fibreglass or similar man made mineral fibre, lightweight fabrication. Larger tanks are likely to be made of aluminium using the stressed skin construction similar to the fuselage.

The normal procedure when flying is to use the fuel in these tanks first so that should any emergency arise these tanks (empty) can be jettisoned as an emergency procedure.

The materials used in the construction of fuel lines vary from one manufacturer to another and in the type of aircraft in which they are fitted. They may consist of stainless steel, aluminium alloy, flexible rubber or reinforced neoprene rubber hose. The diameter of a fuel line is consistent with the size of aircraft ranging from 3mm on the smallest to 100mm on the largest aircraft; the average being approximately 10mm–15mm. **Great care must be taken if cutting has to be carried out as fuel lines are routed through a great many structural elements such as the cabin floor, cargo holds and many other locations.**

1.5 Powered and Pressurised Systems

The internal spaces in an airframe are closely packed with components, pipes, cables and ducts of various different systems. In large aircraft most of these systems are run through the cargo holds but pressurised containers and pipes can be found in many locations on modern aircraft. These systems have the effect of aggravating a fire situation and even impeding the penetration of fire fighting media. Typical pressurised and powered systems found on aircraft include:

1.5.1 Hydraulic and de-icing systems

Hydraulic (pressurised liquid) systems are normally used to operate undercarriage assemblies, flaps and brakes. The hydraulic fluid may contain a castor oil/alcohol mixture, or certain mineral oils or even synthetic liquids. The de-icing systems are normally non-pressurised and generally alcohol based. The capacity of these systems range from 5–225 litres.

Hydraulic systems can be pressurised up to 70 bars. Any fracture or rupture of a system will release fluid in mist form, which is readily ignitable and causes acute irritation to the eyes.

1.5.2   Electrical systems

The electrical systems on modern aircraft are used to operate a multitude of varying devices, from radar and navigational aids to hot plates within the galley. Batteries or engine driven alternators with rectification to direct current power these systems. Aircraft engines may have their own battery system or the aircraft may have a central battery position to give a reserve, peak or start-up loads.

It is important for firefighters to take care when dealing with these systems as circuits may have been damaged or even cut and any operation of battery isolation switches may result in a spark of sufficient power to ignite any flammable vapour present.

1.5.3   Auxiliary Power Units (APU)

Many modern aircraft are equipped with an auxiliary generator powered by a small turbine engine. It is often found in the tail cone area or to the rear of the aircraft. It is normally used on the ground to run various services whilst the main engines are shut down. Its operation also restores battery levels.

1.5.4   Pressurisation and air conditioning

The fuselage of a pressurised aircraft is specially strengthened and each door is pneumatically sealed. The pressurisation and air conditioning is effected by the engine driven compressors. The pressure within the fuselage is maintained at approximately 1 bar throughout the flight until the aircraft nears the ground. Automatic vents come into operation so as to provide a gradual equalisation of any difference between the internal and external pressures due to local weather conditions.

1.5.5   Compressed gases

There will be a variety of compressed gases present on an aircraft such as, Compressed Air, Nitrogen and Oxygen. These have varying uses from life saving, to pressurisation of fuel and hydraulic systems in an emergency.

Identification marking of compressed gas containers can be found in Chapter 2, Section 2.4.2.

1.6   Seating

The internal layout and seating arrangements of an aircraft will vary widely and generally reflect the type of flight (i.e., scheduled, chartered, cargo or combination), destination, and the aircraft type. Contemporary passenger carrying aircraft are able to quickly vary seating arrangements and passenger carrying capacity using first, business and economy classes in determining cabin density. The seating and its adjustment, together with seat belts are important aspects in the rescue of aircraft occupants in any accident. Firefighters should familiarise themselves with the different type of seat and seat belt whenever they have the opportunity to do so. Seat construction has changed over the years from those of small tubular frames to composite one-piece moulded armchairs. The latter may pose difficulties should it need to be cut. Examples of seating arrangements are shown in Figures 1.6 and 1.7 but are by no means standard.

Another important factor in rescue operations is the spacing between seat rows known as the pitch. This can be as little as 740mm with the main aisle gangway further confined in some chartered holiday flights of high passenger density. This can create difficulties with access and in setting up rescue operations. A typical cross section of a civil passenger aircraft showing the width restriction in the aisle gangway is shown in Figure 1.8.

The seating arrangements and passenger carrying capacity of aircraft of similar types will vary considerably between different carriers and the type of operation the airline is engaged upon.

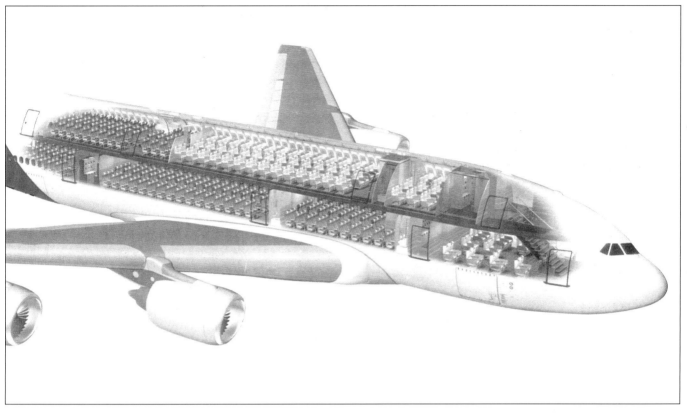

Figure 1.6 Future trends are leading to two wide-bodied decks. (Diagram courtesy British Aerospace Airbus)

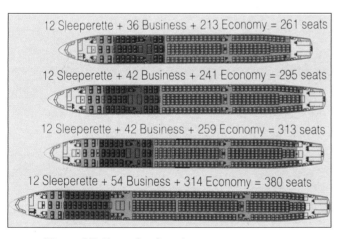

12 Sleeperette + 36 Business + 213 Economy = 261 seats

12 Sleeperette + 42 Business + 241 Economy = 295 seats

12 Sleeperette + 42 Business + 259 Economy = 313 seats

12 Sleeperette + 54 Business + 314 Economy = 380 seats

Figure 1.7 Example of seating arrangements on a modern aircraft. These arrangements are by no means standard and will reflect the type of flight, i.e., scheduled, chartered or combination.

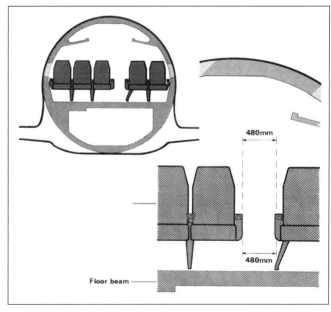

480mm

480mm

Floor beam

Figure 1.8 Illustrates the aisle gangway restriction which may be present on some chartered flights.

1.7 Aircraft access and exits

1.7.1 Doors

The normal means of entrance into aircraft is through a door. The number and position of doors differ greatly between aircraft and operators. These differences may extend to the same type of aircraft with different operators. Consideration should be given to:

● The age of the aircraft

● The size of the aircraft

● The type of aircraft

Aircraft are provided with at least one normal entrance or exit in the form of a main door, with most large passenger aircraft having several others. Normally the main door is on the left side of the aircraft, bearing the number 1, with others consecutively numbered aft. The same arrangement is found on the right side, numbering from forward to aft. In an emergency all doors and other escape routes, such as windows and hatches, will become available to facilitate the rapid evacuation of the aircrafts' occupants. (See Figure 1.9.)

The doors and exits are normally of a size and number relating to the passenger carrying capacity of the aircraft.

The location and operation of an aircraft door is fundamental to any rescue of an aircraft's passengers and crew. Liaison visits to airports in conjunction with the airport fire service should facilitate this operational rescue element being incorporated into ongoing training programmes.

Aircraft doors are readily identifiable, being outlined in a contrasting colour. Where cabins are divided into two or three compartments there will be at least one exit door in each unless there is ready and easy access from one compartment to another. Doors are designed to operate in a variety of ways although the most common encountered on modern passenger carrying aircraft is the "plug" type. When "plug" doors are opened, they initially move inwards toward the cabin breaking the seal and then are pulled out or may retract into the ceiling void. This type of door will not open if the cabin remains pressurised. Other doors may be hinged on the forward side and open outward; some pull out and move sideways. The operation of control handles or switches are clearly marked on both the inside and outside of the aircraft.

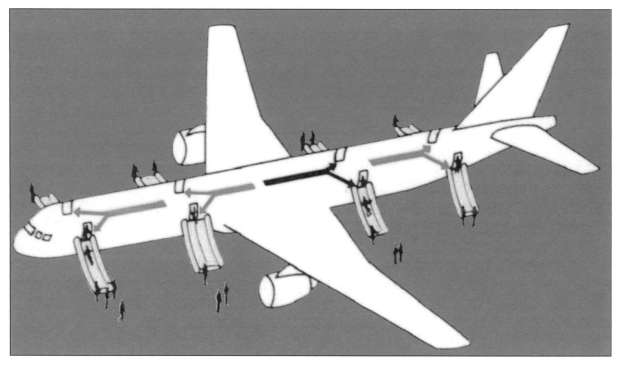

Figure 1.9 Illustration of an emergency evacuation on a Boeing 757 aircraft with all escape slides deployed.

Figure 1.10   Aircraft evacuated using escape slides.

Figure 1.11   Note over wing deployment of centre chute.

1.7.2 Emergency stairs

There are some aircraft still operating that have a rear entrance fitted with stairs for normal embarking and disembarking e.g., DC9, BAC 111. These stairs are usually lowered by hydraulics, but they do have a manual override facility for use in an emergency. Operating instructions are clearly marked on the outside of the aircraft. (See Figure 1.12.)

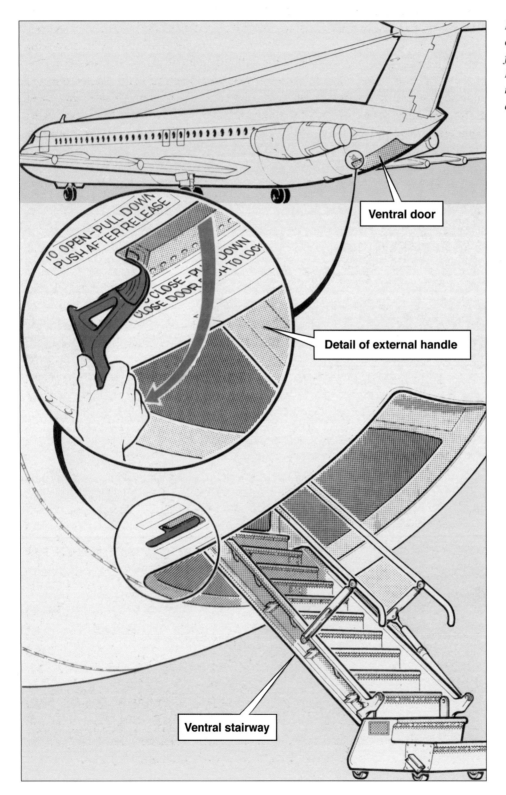

Figure 1.12 Example of a ventral stairway found on a DC9 and BAC 111 and external manual operation in an emergency.

Ventral door

Detail of external handle

Ventral stairway

1.7.3　Escape chutes or slides

These devices are provided for the rapid evacuation of passengers in an emergency. (Figure 1.13.)

They are predominantly inflatable being self supporting and fabricated from nylon and rubber. The slides are inflated by the operation of nitrogen or carbon dioxide cylinders which in turn are coupled to the door opening mechanism.

On older aircraft types non inflatable escape slides may be found. These are strongly constructed of nylon or other synthetic materials and provided with hand-holds. Whilst this type of slide is effective the requirement to support it whilst in use has led to its demise on all contemporary passenger carrying aircraft.

In most modern aircraft the slides are deployed automatically when the door is opened in an emergency from the inside. Should the door be opened from the outside the mechanism used in deploying the slide will automatically disarm. **However, fire-fighters should act cautiously when gaining entry particularly if the aircraft door has been distorted or stressed where the disarming mechanism may be liable to malfunction. In**

Figure 1.13 Illustrating the use of inflatable escape slides.The requirement to support non inflatable slides has led to their demise on all contemporary passenger carrying aircraft.

Figure 1.14 Access being gained following slide removal.

Figure 1.15 Scene following emergency evacuation.

some circumstances an escape slide may accidentally inflate whilst still inside the aircraft impeding any evacuation and/or access by firefighters. (See Chapter 7, Section 7.2.1 (d).)

It should be noted that some passengers, in using the emergency chutes/slides, may incur minor injuries. These injuries will usually range from sprains, minor friction burns to broken bones (ankles).

1.7.4 Windows

These are usually completely fixed, often double glazed and made of transparent plastic. In modern pressurised aircraft the windows are kept as small as possible, are extremely strong, and in very tough frames. **They should, for rescue purposes, be avoided as far as possible.**

1.7.5 Emergency Hatches

Emergency hatches are fitted to most civilian aircraft in the shape of window panels, these are designed to fall inwards or outwards on the operation of a self release mechanism.

These types of hatches would normally be found as part of the emergency exits over the wings. They are primarily used for escape routes in case the aircraft ditches in the water. They can also be found on the top of the fuselage.

These exits can be easily operated from both outside and inside the aircraft.

1.7.6 Break-in points

These are areas marked on the fuselage by broken lines (see Figure 1.16) where it should be possible to cut through the airframe to force an entry. They are usually located where there are no internal obstructions such as electrical wiring or pipework. They are not weak points in the airframe structure but merely areas between frames. Only the skin and stringers should therefore require cutting, but owing to the arrangement of internal fittings, these points are usually located well up on the fuselage and well above the aircraft's cabin floor. This can cause problems with access and requires a suitable working platform should forced entry be attempted. **All other options in gaining entry to an aircraft should be attempted before resorting to cutting in.**

1.7.7 Other factors

(a) Cargo Aircraft

Although firefighters are primarily concerned with the rescue of an aircraft's occupants, they may be called upon to tackle an incident involving a cargo aircraft. Obviously the number of people involved will be relatively few, probably confined to the crew alone. There may be animals on board, however, domestic livestock such as horses or cattle or indeed wild animals being either imported or exported. It is important that the Incident

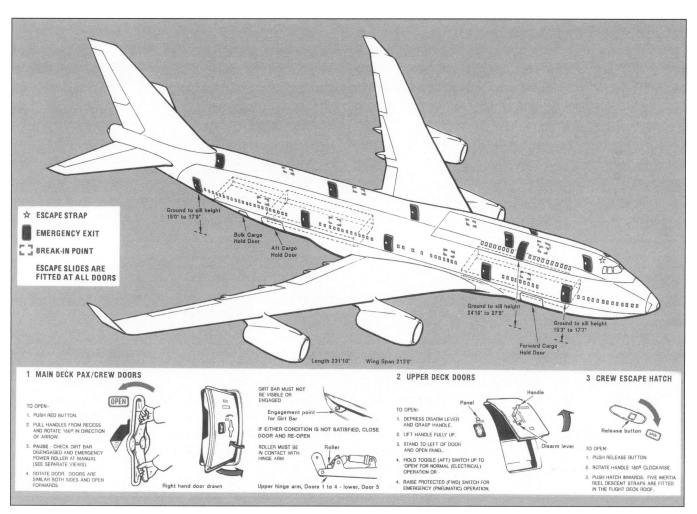

Figure 1.16 Emergency access chart – Boeing 747 *(Diagram courtesy British Airways)*

Commander determines at the outset what cargo the aircraft is carrying. (See Figures 8.2 – 8.5.) This information may normally be gained from the crew before resorting to the examination of documentation which may delay operations. The means of access to a cargo aircraft will differ from a passenger carrying aircraft although there will invariably be a main door. This will allow access to the main cabin area but not generally the freight holds. These may usually only be entered via the cargo doors. These doors are often fitted with pressure equalising panels which generally require power to operate them. The A.P.U. or the aircraft's engines must, therefore, be run unless sufficient battery power is available to operate them.

(b)   Depressurisation

Aircraft are equipped with means of pressurising and depressurising the fuselage. Should the automatic depressurisation vents fail to operate there may be a difference between the inside and outside pressure of the aircraft when it lands. If the aircraft cabin remains pressurised to any extent, (as little as 103 pascals) entry by the main cabin doors, will be unachievable. In this case the fuselage skin has to be penetrated in order to equalise the pressure.

1.8   Fire

Most modern aircraft have fire detection and extinguishing systems. They are usually independent of each other in that detection is automatic but an extinguishing system operates only when it is switched on by the flight crew. The extinguishing agents used are BCF and BTM or freon combined with nitrogen contained in pressurised cylinders located in various parts of the aircraft. (Given a moratorium on the production of ozone depleting extinguishing agents, others are actively being sourced).

These systems are only intended to cope with a wholly enclosed fire in a particular compartment, e.g., engine, luggage, fuel and have no effect outside the confines of that particular compartment. In an accident the extinguishant may disperse given damage to a compartment or be overwhelmed by any large fuel fire. If cylinders are not damaged and are involved in fire they may burst at a comparatively early stage.

1.9   Flight Recorders

Flight recorders are carried on all civil passenger carrying aircraft and will also be found on many aircraft used for freight or private or business purposes. Although still referred to in the media as "black boxes", they are invariably brilliantly coloured fluorescent red. Usually found in the rear fuselage section of an aircraft, adjacent a rear exit door, (see Figure 1.17) they are designed to withstand shock, fire and water ingress. They are by no means indestructible however, hence their positioning aft where any forward impact will be absorbed by the airframe structure. There are usually two flight recorders, a flight data recorded (FDR) and a cockpit voice recorder (CVR), usually mounted together although this is not always the case. These record the flight details of the aircraft and any communication or flight deck activity. The flight recorders are invaluable to the Air Accident Investigations Branch (AAIB) (see Chapter 6, Section 6.8.6) and should not be handled unless absolutely necessary or they are in danger of being irretrievably lost.

Figure 1.17 A flight data recorder (FDR) and a cockpit voice recorder mounted in an aircraft.

Aircraft Incidents

Chapter 2 – Military Aircraft

2.1 General

Fixed wing military aircraft range from very large four engine transports to single seat fighters, including supersonic attack aircraft and those capable of vertical take-off and landing (VTOL). Military aircraft often crash a long way from their airfield bases and therefore the likelihood that local authority firefighters may have to carry out initial rescue and firefighting is a distinct possibility. There is also the possibility that military aircraft may use a civil airport for a number of reasons including air displays, diversions, training flights, troop movements or temporary stationing. With this in mind, some knowledge of the construction design, special features and rescue techniques on contemporary military aircraft is essential for firefighters. It is not possible to cover every type of military aircraft operating as types vary within the Royal Air Force and the Royal Navy, the United States Air Force and other military aircraft either based or visiting the UK. The foremost military aircraft in use today within the United Kingdom and Europe are the Tornado and Eurofighter Typhoon, the Harrier jump-jet (VTOL) together with a variety of larger aircraft performing such tasks as troop transport, refuellers and cargo transport. Smaller military aircraft such as the Hawk and Tucano are used as pilot trainers.

2.2 Construction

The fundamental points made in Chapter 1 regarding the construction of civil aircraft apply equally to similar aircraft in military service used for air freight or troop transport roles and to military aircraft generally.

Some features of military aircraft do however differ from their civilian counterparts e.g., the increased use of composite materials in the construction of fighter aircraft, reinforced flooring, side seating, loading ramps and weapons. The features of most interest to firefighters are set out in this chapter with the exception of weapons systems which are dealt with in Chapter 8, Special Hazards.

Firefighters should take every opportunity to liaise with RAF, RN and USAF fire departments visiting any military airfields in their area. Information from such sources may prove invaluable at any military aircraft accident/incident.

2.3 Fuel and Fuel Tanks

In addition to the aviation fuels listed in Chapter 1, section 4, some military aircraft, such as the NATO F16, use hydrazine as a fuel supply for the emergency power unit (EPU). Hydrazine comprises 70% hydrazine and 30% water, however, it should be treated as pure hydrazine when dealing with the substance. **Hydrazine is a "Hypergolic" fuel, a substance that will spontaneously ignite in the presence of an oxidising agent.**

Physical Properties:

- It is a clear oily liquid with a distinctive odour similar to ammonia
- It has a flash point of 60–90°C
- It has an auto-ignition temperature of 270°C
- It has a wide range of flammability (47% upwards by volume with air) therefore:
  (a) only a small amount is required to create a flammable atmosphere
  (b) it is unlikely to create "Too rich a mixture" in confined spaces
- It is corrosive
- It can self ignite (hypergolic)
- Its vapours present explosive hazards

- It is insensitive to shock and therefore very stable
- It is miscible with water (it can be diluted) this must be followed by a neutralising agent.
- Special military teams are required to remove the material for disposal.

Whilst hydrazine is an unusual and volatile fuel it is carried in **very small quantities** of some 20 to 25 litres.

Always wear liquid-tight chemical protective clothing including breathing apparatus when dealing with hydrazine emergencies. It can be absorbed through the skin and have serious effects on the nervous and respiratory systems. Exposure to the liquid or vapour should be kept to a minimum.

Fuel Tanks

The fuel tank systems on military aircraft are similar to those fitted in civil aircraft. However, smaller scale tankage will be found on fighter and trainer aircraft. (See Figure 2.1.)

Where military aircraft are required to undertake long range sorties, they will usually be equipped with additional fuel tanks, these can be either internal or external. Wingtip tanks, ventral fuselage tanks and wing drop tanks are the standard types. Larger military aircraft may have temporary fuel tanks fitted within their fuselages.

Large military aircraft, such as the VC10 and the Tristar (1011) are used for in-flight refuelling and are specially designed internally to carry several tonnes of fuel.

2.4 Powered and Pressurised Systems

2.4.1 Auxiliary power units

An auxiliary power unit (see Chapter 1, Section 1.5.3) is installed in most military aircraft. It is not essential for propulsion but to provide a supplementary source of electrical or pneumatic power, e.g., to drive an additional alternator, to pressurise pneumatic turbo-starters to the main engines, or to supply pressurised air for a nuclear weapon. It normally has its own compartment close to the main engines with its own intake and exhaust and uses various types of aviation fuel.

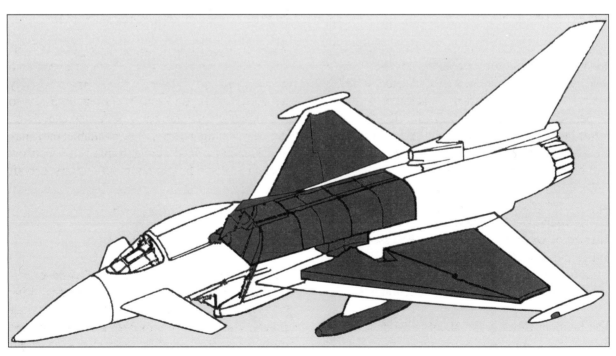

Figure 2.1 An example of a fuel tank system on a military aircraft (Euro-Fighter EF 2000).

2.4.2 Liquid Oxygen and Gases

Military aircraft carry quantities of liquefied oxygen (LOX). This supplies, via converters and other equipment, oxygen to personnel on board. The systems are usually located near the cockpit, but additional containers on transport aircraft may be located at the sides or rear of passenger compartments.

Systems using containers of gaseous oxygen or other gases may also be found within the aircraft and can be identified by their colour (they will also have the chemical formula and lettering displayed around the neck of the cylinder).

Most common in use are:

● Oxygen (black)
● Nitrogen (light grey with black neck)
● Carbon Dioxide (white)
● Air (light grey)

All liquid oxygen and gas containers are pressurised and may therefore explode or burst violently when involved in a fire.

2.5 Rescue from Military Aircraft

2.5.1 General

Firefighters should remember that a call to a crashed military aircraft may not necessarily involve rescues. The crew may have ejected before the crash and the aircraft may therefore be empty of personnel. A high speed crash will invariably result in the disintegration of the aircraft and if the crew have not baled out, their chances of survival will be minimal.

In cases where rescue and firefighting is demanded, firefighters should be careful to avoid passing in front of weapon racks, and in the case of large military transport aircraft, they should assume the aircraft may be carrying weapons unless and until informed otherwise. (Chapter 8, Section 8.3.)

The Incident Commander should also be aware of the potential initiation hazard radio transmissions pose with military ordnance. Current guidance is

that no radio transmitting equipment should be used within 10 metres of the risk. Vehicles fitted with mobile radio terminals should not be taken within 50 metres of the risk unless the radio is switched off. (see Fire Service Manual, Volume 1, Communications and Mobilising).

The rescue of aircrew may be arranged in four consecutive phases:

● Gaining access
● Making the ejector seat safe
● Releasing the aircrew from seat and harness
● Lifting out and removal to safety

2.5.2 Gaining Access

Doors

Firefighters should be aware that doors on military aircraft may differ from those on civil aircraft. On large aircraft there is usually at least one main door with operating instructions stencilled on it. This is usually found on the left side of the aircraft but may be found under the nose. Doors may open inwards, outwards, slide up or down on hinges or in some cases, can be entirely jettisoned. All are finely balanced so that unless there is extreme distortion one person should be able to open them. Instructions on the fuselage are usually sufficiently explicit.

Occasionally, large military transport aircraft have freight doors similar to those on civil cargo carrying aircraft. These doors are normally hydraulically operated which will require power to be available to the aircraft in order to open them. Should hydraulic power not be available, they may be manually cranked although this is a very slow procedure. Some freight doors on military aircraft may, indeed, only be opened from inside and therefore rescuers must gain access elsewhere.

2.5.3 Emergency Hatches

The emergency hatches in military passenger aircraft are usually in the form of removable windows, these are similar to those found on civilian aircraft. These areas are usually delineated on the fuselage and the operating instructions stencilled adjacent.

2.5.4 Passenger seats

In some fixed-wing military transport aircraft such as the VC10, the passenger seating faces rearward and is grouped similarly to that of civilian airliners although usually with a narrower aisle. The seating is easily removable for conversion to freight carrying. Seating in mixed freight and passenger loads can be found either in the front or rear fuselage.

Paratroop seating is very basic, consisting of tubular frames with straps of webbing or synthetic material. These are set in rows along the sides of the aircraft and, in some cases, also back-to-back down the centre. The maximum number of paratroopers which can be carried is greater than the number of ordinary seated passengers.

2.5.5 Break-in Points

All large military aircraft have break-in points similar to those found on civil aircraft. (See Chapter 1, Section 1.7.6.) They should be regarded as a last resort, to be used only if all other means of access is restricted.

2.5.6 Cockpit Canopies

Cockpit canopies are found on military fighter aircraft. They are usually slightly domed hinged or sliding perspex hoods over the seat positions behind a fixed windscreen of laminated safety glass or perspex being separate from the canopy.

Sliding Canopies

There are internal and external canopy release handles which, when operated, open the canopy locks, enabling the whole canopy to be removed. Appropriate instructions are stencilled on the fuselage. Most canopies are quite heavy and should be handled accordingly.

Hinged Canopies

These canopies usually have a jettison mechanism which can be operated internally or externally by means of a canopy jettison release. Appropriate instructions are stencilled on the fuselage. The jettison release opens the canopy locks and fires a cartridge which pushes the forward section of the canopy upwards pivoting it on the rear or side hinges and throwing it back or sideways. If the aircraft were in flight the slipstream would tear it clear of the aircraft immediately. The jettison mechanism operates automatically should the aircrew eject.

When the aircraft is on the ground the canopy may be exploded clear of the aircraft or it may merely roll towards one side or another or just lift slightly and therefore may require to be manhandled clear. The number of personnel engaged in such operations should be kept to a minimum and all other personnel and vehicles not actively involved should be removed to a safe distance from the aircraft whilst these operations are taking place.

2.5.7 Miniature Detonating Cord (MDC)

Some aircraft have a miniature detonating cord (MDC) built into the perspex of the canopy (See Figure 2.2.)

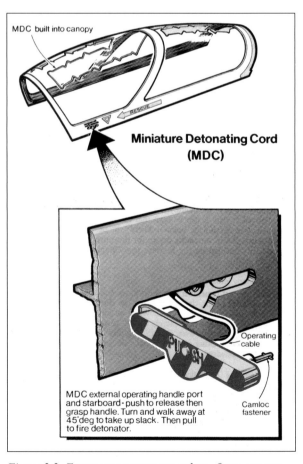

Figure 2.2 Emergency rescue procedure. Open canopy manually or fire MDC

This device is potentially very dangerous being able to project fragments of the canopy for up to 20m. Firefighters should therefore only operate it if the normal release mechanism is inoperable. The following procedures should be adopted:

i.   A firefighter should approach the cockpit in full view of the aircrew and be prepared to act on any signal from the aircrew that they are about to operate the MDC, e.g., turn away, crouch down as low as possible. A flash hood should be worn, plus helmet visor in the down position.

ii.   The firefighter should remove the handle, as per the instructions on the fuselage, extend the cable and move to the front of the cockpit as far as the cable allows. He should then face away from the cockpit, lower his head and give the cable a sharp pull. (See Figure 2 .3.)

Firefighters must take extra care if the aircraft is nose down, or canted towards the MDC operating gear side, since this could bring the trajectory of the fragments down to a low level. It may well be considered prudent to tie a line on to the MDC handle to extend the pull from a safer distance.

Involvement in fire could cause the MDC to detonate and an MDC equipped canopy that has become detached from the aircraft could prove to be lethal and **should not be approached or touched unless absolutely necessary**.

There are two other points which firefighters should bear in mind. Firstly, if the crew are still in the aircraft they may be disorientated and a cautious approach may be necessary in case they operate the MDC without realising rescue is at hand. Secondly, the engines may still be running, in which case firefighters should be careful to stay clear of the intakes and exhausts when approaching the aircraft.

2.5.8   Breaking-in of Canopies

Attempting to smash the canopy with an axe or other such tool is absolutely the last resort. Perspex is very strong anyway and in modern aircraft it is laminated and virtually unbreakable. Apart from the hazards to the aircrew, there is always the

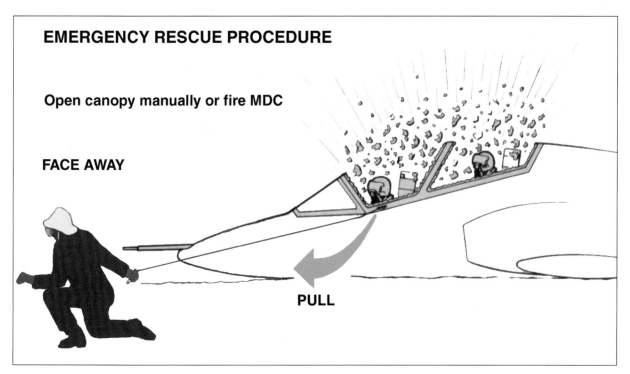

EMERGENCY RESCUE PROCEDURE

Open canopy manually or fire MDC

FACE AWAY

PULL

Figure 2.3 Manual operation of the M.D.C.

chance of setting off the MDC or ejection seat with disastrous consequences. Depending on the type of tool carried, it might be possible to cut the canopy with certain types of rotating saw, but the same hazards could apply.

2.6 Making Ejection Seat Safe

All types of high-speed combat aircraft are fitted with ejection seats which are designed to 'fire' the crew from the aircraft in an emergency. (See Figure 2.4.)

The types of seat used in the RAF, RN and USAF are similar in that they all have an explosive charge or rocket pack to eject the crew member upwards. They do, however, differ in a number of details, and these differences are important to the safety of firefighters and aircrew. There is, moreover, the possibility of frequent visits by military aircraft from other NATO countries.

It is beyond the scope of this book to describe all types of ejection seat in use today, and indeed, technology will enhance and advance their design in the future. The fundamental considerations in making a seat safe are however outlined. Officers who may be required to attend military aircraft accidents/incidents should take every opportunity of familiarising themselves and their personnel with as many types of ejection seat as possible.

2.6.1 Martin Baker Ejection Seats

Almost all Martin Baker ejection seats can be made safe by the insertion of the safety pin into the hole in the sear behind the head of the seat. (See Figures 2.5 and 2.6.) The only exception at present is the Tornado aircraft where there is no sear at the head but the pin is inserted in between the legs in the seat pan firing handle.

The safety pin stowage may vary between different aircraft, but in general firefighters should look for **bright red or orange coloured tabs** and from these select the "main gun sear pin".

Once the main gun sear pin has been inserted, rescue may commence as circumstances dictate. The remaining sears can be similarly made safe if practicable, given the urgency of the release of the aircrew.

Figure 2.4 Ejection seat operation and symbol.

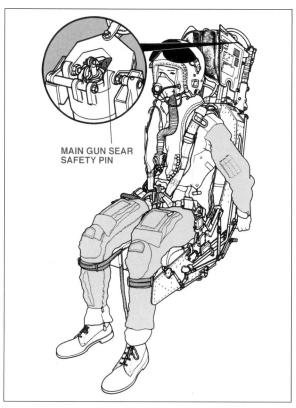

Figure 2.5 Ejection seat safety pin installed – typical. (Harrier T Mk 8)

2.6.2   US Ejection Equipment

There are a considerable variety of US aircraft flying in the UK, and, consequently, a number of different types of ejection equipment. Martin-Baker seats are fitted to some US aircraft, similar to those in British aircraft, but there are also many US aircraft with US designed seats of various types. There are even different marks of ejection seats for the same type of aircraft, e.g., an F15A may be differently equipped to a F15D.

US aircraft have instructions stencilled on the fuselage on how to open canopies. Inside the cockpit, the ejection seat firing handles are easily detected by their **black and yellow striped colouring (similar to those in UK and NATO aircraft)**. The locking pins for the handles and sears are however, not kept in racks as in British aircraft but in nylon bags at the side, or more generally behind the aircrew. The pins are of different sizes to obviate the danger of being put in the wrong sear and can often only be inserted one way. They may also be of a type with a press-button top which has to be depressed to get the pin in the sear.

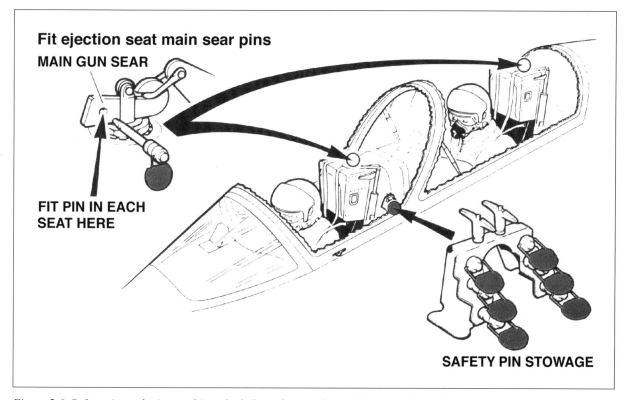

Figure 2.6 Safety pins to be inserted into the hole in the sear behind the head of the pilot and crew. Safety pin stowage may vary. Firefighters should look for bright red or orange coloured tabs and from these select a main gun sear.

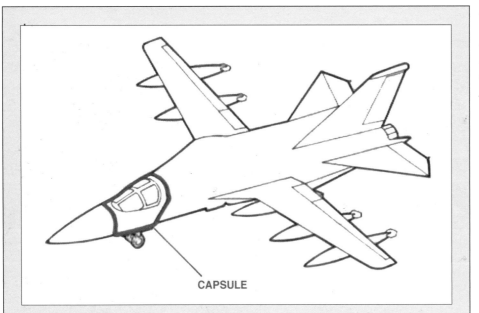

Figure 2.7 Diagram of the ejection capsule of a USAF F11 aircraft. Similar instructions are stencilled on the fuselage.

F-111   Aircraft entry (all models)

1   Emergency/Normal Entry

(a) Push internal lock release button, located both sides of aircraft below canopy rails.

(b) Push forward end of external handle, located aft of internal lock release button, pull external handle and raise. canopy to locked position.

(c) Position handle back to the lock detent (midway) position to lock canopy open.

2   Cut-In

(a) Cut canopy along canopy frame. CAUTION – DO NOT CUT CANOPY FRAME

Firefighters should, if possible, take time and care to study the cockpit layout and position of the crew before inserting the pins. **Generally, as long as the main ejection seat handles are made safe, as in British aircraft, the remaining pins can be inserted but are not absolutely essential.**

2.6.3   Ejection Capsule

One US aircraft, the F111, has a different ejection system. On this type of aircraft the whole capsule, containing the two aircrew complete with seat and cockpit is fired clear of the aircraft. (See Figure 2.7.) As instructed in Figure 2.7, entry into the capsule is simple and the canopy can be locked up by placing the handle at an angle of 45° to the left. The two ejection handles are located between the aircrew, the pins are kept in a bag behind the pilot who sits on the left side.

The pins are push button type and should be inserted in the top of the handles from the inside out. Between the crew and above are three other sears which can be made safe, if circumstances allow.

Whenever possible firefighters should take time and care studying cockpit layout and position(s) of aircrew before inserting safety pins. Generally once the main gun sear pin is inserted rescue may commence as circumstances dictate.

2.7 Release and removal of Aircrew

UK Aircraft

In modern military aircraft the aircrew are securely strapped into their seats by various harnesses and connections, even to the extent of leg restraints to prevent accidents during ejection. Firefighters should, if possible, take time to read and understand any instructions visible on the aircraft, but it is important that they should know in advance what basic steps need to be taken in order to get aircrew out safely and relatively easily. These steps are listed below.

It is assumed that the canopy, if any, has been removed. Provided that the initial operation in the list is *always* carried out first, the sequence of the remainder is not important and should cause no delay.

(i) **Make the seat(s) safe.** The procedures described in Section 6 should be strictly followed.

Figure 2.8 Remove oxygen mask

Figure 2.9 Release personal equipment connectors.

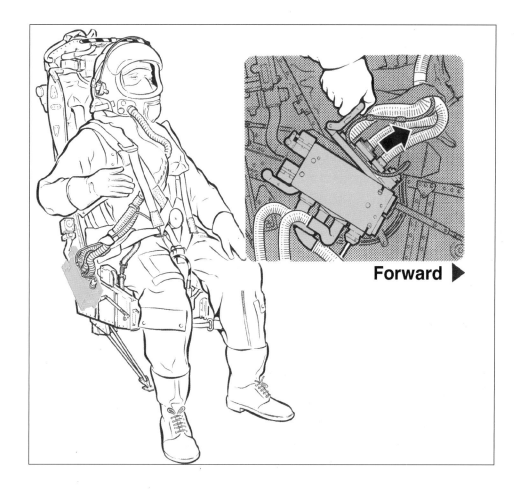

Forward ▶

(ii) Remove the oxygen mask (Figure 2.8) but leave the helmet on if head injuries are suspected.

(iii) Release the personal equipment connector(s) (PEC) by pressing the thumb release and raising the handle. (See Figure 2.9.) PECs vary in size and location but are easily identified. Their release may, in some cases, automatically disengage the leg restraints.

(iv) Release the survival pack connection. (See **2**, PSP, Figure 2.10).

(v) Release the combined parachute and seat harness (Figure 2.11) by turning clockwise and carefully pressing inwards. Firefighters should bear in mind the possibility of the aircrew having internal injuries, and if possible place one hand behind the quick release box.

(vi) If the leg restraints are not released automatically (see (iii)), there may be a small lever at the side of the seat which can be lifted to release them, or the straps may have box clips similar to (iv), (Figure 2.12). Failing this, they will have to be cut.

2.8 Lifting out and removal to safety

While the harness is being removed, the occupant of the seat will usually need to be supported, especially if the aircraft is upside down. There is very little room in the cockpit, so aircrew will have to be lifted from above if the aircraft is in an upright position. The firefighter should straddle the cockpit and attempt to lift the aircrew up sufficiently to lean him over the cockpit sill. Then, with the assistance of other firefighters, the person can be eased onto a stretcher and carried away.

Firefighters should be particularly careful not to operate switches, levers etc., as these could actuate weapons.

US Aircraft

After the seats have been made safe, there are usually two connections to disengage:

(i) a pull-off oxygen-pipe connection; and,

(ii) a harness knob to be rotated, whereby the harness straps are released. The oxygen mask should be removed from the aircrew's face to assist breathing. To lift the crew out he/she should be moved forward, a foot placed behind him on the seat and the body hoisted out back downwards. Firefighters must remember the possibility of back injuries and take appropriate action, especially if they find an ejected capsule of an F111.

2.9 Casualty Handling

The number of casualties in a military aircraft accident will not normally be on the same scale as that associated with large passenger carrying civil aircraft. However, there is every possibility of ground casualties especially where an accident occurs in a built up area. Advice on the treatment of casualties after rescue is given in Chapter 7 Section 4 with additional guidance outlined in Chapter 6, Section 6.6.2.

2.10 Military Aircraft Hazards Database

The Royal Air Force's Aeronautical Rescue Co-Ordination Centre (ARCC) holds a copy of a database on the hazardous material found on military aircraft. This could provide important information to anyone dealing with the aftermath of a military aircraft crash.

In the event of an actual or suspected military aircraft crash, Brigades should contact the Royal Air Force Aeronautical Rescue Co-Ordination Centre (ARCC) which is available on a 24 hours basis.

ARCC will be able to provide information on the hazardous material found on military aircraft and the level of precaution required. This information will be supplied on the clear understanding that neither the Royal Air Force nor any of its personnel will be liable for damage, loss or injury resulting from action taken in reliance to information given. The telephone number (provided to all brigades) should also be used to request the use of Royal Air Force helicopters for the movement of offshore fire-fighting teams if required.

Figure 2.10 Air crew release.

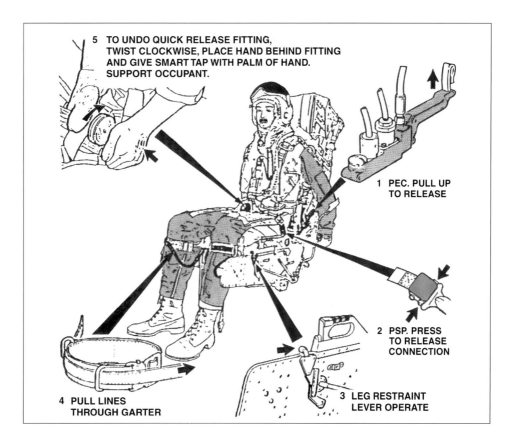

5 TO UNDO QUICK RELEASE FITTING, TWIST CLOCKWISE, PLACE HAND BEHIND FITTING AND GIVE SMART TAP WITH PALM OF HAND. SUPPORT OCCUPANT.

1 PEC. PULL UP TO RELEASE

2 PSP. PRESS TO RELEASE CONNECTION

3 LEG RESTRAINT LEVER OPERATE

4 PULL LINES THROUGH GARTER

Figure 2.11 Combined parachute and seat harness.

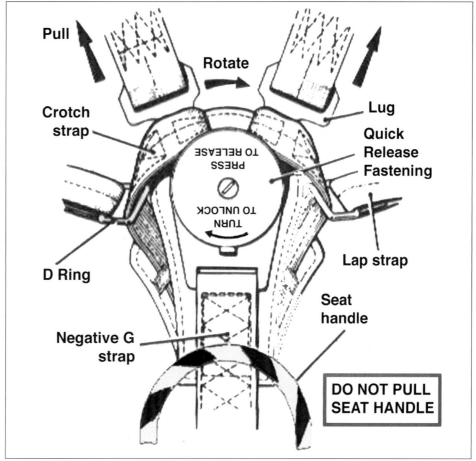

Pull

Rotate

Crotch strap

Lug

Quick Release Fastening

PRESS TO RELEASE

TURN TO UNLOCK

Lap strap

D Ring

Seat handle

Negative G strap

DO NOT PULL SEAT HANDLE

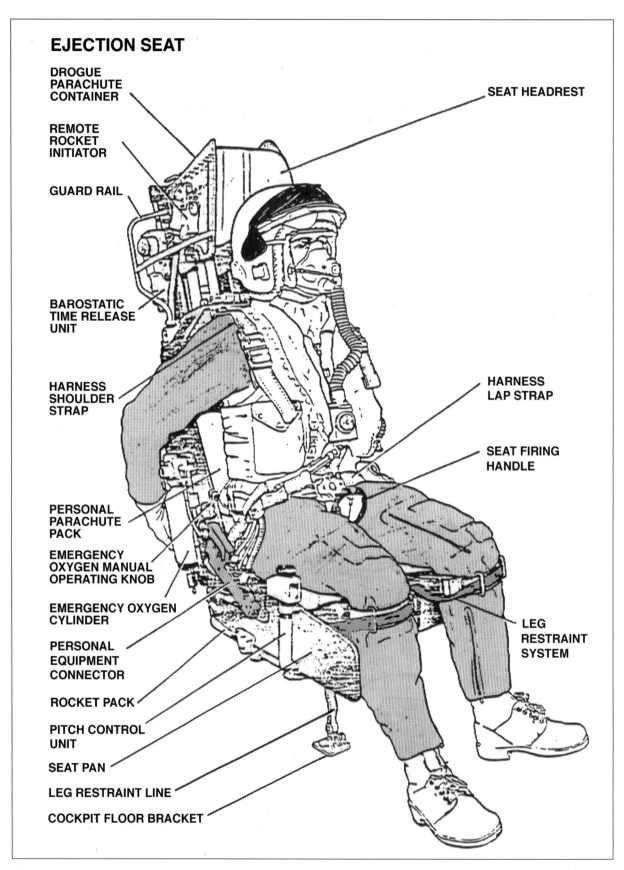

EJECTION SEAT

DROGUE
PARACHUTE
CONTAINER

REMOTE
ROCKET
INITIATOR

GUARD RAIL

BAROSTATIC
TIME RELEASE
UNIT

HARNESS
SHOULDER
STRAP

PERSONAL
PARACHUTE
PACK

EMERGENCY
OXYGEN MANUAL
OPERATING KNOB

EMERGENCY OXYGEN
CYLINDER

PERSONAL
EQUIPMENT
CONNECTOR

ROCKET PACK

PITCH CONTROL
UNIT

SEAT PAN

LEG RESTRAINT LINE

COCKPIT FLOOR BRACKET

SEAT HEADREST

HARNESS
LAP STRAP

SEAT FIRING
HANDLE

LEG
RESTRAINT
SYSTEM

Figure 2.12 Ejection seat layout showing position of various connectors. If leg restraints are not released automatically they may have to be cut.

Chapter 3 – Rotary Wing Aircraft

3.1 General

Rotary-wing aircraft, which are mostly of the helicopter type, have developed very quickly, both technically and commercially. The helicopter is often selected for use because of its capability of gaining access to an area which is inaccessible to other types of aircraft. Over many years the helicopter has proved its versatility in a wide range of roles which include:

- Military assault
- Anti-submarine activities
- Police surveillance

- Fire-fighting
- Medical evacuations
- Search and rescue (both on and off-shore)
- Short range public transport
- Company and private transport

The level of rescue and fire-fighting protection required to be provided for helicopter movements is in line with that for fixed wing aircraft. This considers helicopter sizes, the volume of fuel and passenger carrying capacity.

A study of the characteristics of helicopters has shown that three rescue and fire-fighting cate-

Figure 3.1 GKN Westland EH 101 Heliliner.

gories are adequate to cover the range of helicopter operations. These categories are defined on the basis of the overall length of the helicopter including tail, boom and rotors.

Category	Helicopter Overall Length
H1	Up to but not including 15m
H2	Up to but not including 24m
H3	Up to but not including 35m

In the case of a heliport located on an aerodrome used by fixed wing aircraft, the rescue and firefighting facilities provided will normally be adequate for the protection of helicopters.

As with the categorisation of airports, rescue and firefighting provision at heliports is increased as the category rises.

3.2   Construction

The construction of the airframe of a helicopter is very similar to that of the fuselage of a fixed wing aircraft but of a lighter design. There are several reasons for this:

- The airframe is not stressed to carry a mainplane.
- The cabin is not pressurised for high altitude flight.
- The undercarriage assemblies are relatively small.
- Structural members (although of the same materials) are of a much smaller cross-section.
- Sheet metals used are of a thinner gauge.
- The use of composite materials, such as Carbon Fibre Reinforced Plastic, Aramid Reinforced Plastic and lighter alloy metals, such as Aluminium, Duralumin, Elektron and Alclad magnesium are used extensively.

Figure 3.2 shows the typical areas in a modern helicopter where composite materials are used.

3.3   Engines

The majority of helicopters in use today use the turbo-shaft engine, geared to drive rotor heads instead of propellers, although some of the smaller helicopters still have a piston engine driving the rotors.

The location of the engines will depend on the design of the helicopter. The engines can be found as follows:

- in the nose section of the aircraft
- mounted over the cabin
- pod mounted on the rear pylons (Chinook)

Where the engines are located within nacelles, they may be fitted with fire access panels and are separated from the rest of the aircraft by fire-resisting bulkheads. Where twin engines are mounted side by side these will also be separated by a fire resisting bulkhead.

3.4   Rotors

Most helicopters have one large overhead rotor and a small stabilising rotor at the tail. The only exception, at present, is the Boeing Vertol/Chinook which has two overhead rotors. These are contra-rotating and therefore do not require a stabilising rotor.

The number of rotor blades varies from helicopter to helicopter, but normally it is between 2–5 blades, however there may be more. The diameter of the rotor blades can be as much as 19m.

The rotor blades are constructed from a mixture of both composite materials and light alloy metal. This construction method poses a hazard to firefighters for, should the rotor blade shatter, **it will throw portions of the blade a great distance**. This may also **increase the hazard of inhalation of composite materials**. There are other hazards associated with rotor blades which firefighters should be aware of. These include:

- Helicopter engines may still be running even after a crash and although the main rotor may have shattered, the tail rotor could still be rotating. This hazard is

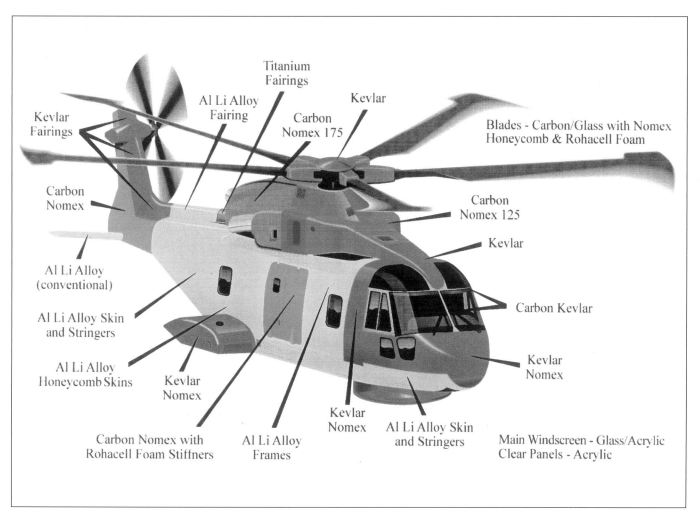

Figure 3.2 Composite materials used in the construction of a modern helicopter.

increased in poor visibility or adverse weather conditions. Tail rotor blades have been coloured in bright designs to improve their visibility when in motion.

- The angle of the crashed aircraft is very important to firefighters. A nose down situation will bring the rotor tips very close to the ground.
- On some helicopters as the rotors slow down they droop and can fall dangerously low.
- On the Boeing Vertol/Chinook the front of the helicopter is already canted forward and down and when slowing this aircraft's rotor tips fall very low.

The best approach to the aircraft is from the front in full view of the pilot. (See Figure 3.3.) If the aircraft has crashed, approach from the rear along the tail boom on the opposite side to the tail rotor keeping as low as possible to avoid the main rotor. Firefighters should approach these aircraft with extreme caution.

3.5 Seating

The seating arrangements will vary according to the design and size of the aircraft, but in the larger multi-seat types, it is similar to that of small fixed wing passenger aircraft. Seats are usually of the metal tubular frame type or of a one-piece moulded composite material. The internal layout of the seats again will vary between operating companies, but in general, are arranged in rows of four across the cabin or rows of two and two with an aisle or gangway in between. All seats are fitted with seat belts. (See Figure 3.4.)

HELICOPTER DANGER AREAS

ONLY ENTER THIS AREA WITH THE PILOT'S PERMISSION

SAFE TO ENTER ONLY WHEN CLEARED

DANGER – NEVER ENTER

Figure 3.3 The best approach to a helicopter is from the front in full view of the pilot.

3.6 Access and Exits

3.6.1 Doors

Doors are usually hinged or side-sliding, with a variety of opening devices. In some aircraft the door may be jettisoned from both inside and outside the aircraft. Firefighters must be prepared to take the weight of these doors as they come away. The Boeing Vertol/Chinook has some unusual features, as some of the doors separate into sections, the lower part of the door folding out to form steps, and the upper sliding up, in and over. (Figure 3.5)

As with fixed wing aircraft the operating instructions for all doors are marked on the outside.

3.6.2 Escape Panels

The larger helicopters usually have, in addition to the normal doors, several areas of the fuselage which can be removed or cut into in an emergency. They may be specially marked sections around certain areas prominently indicated on the fuselage where cutting-in can take place. The Boeing Vertol/Chinook also has a cargo door escape panel at the rear and if necessary, once firefighters are inside, the windows can be kicked out.

On small helicopters, if for any reason the doors are jammed, the construction is relatively light and

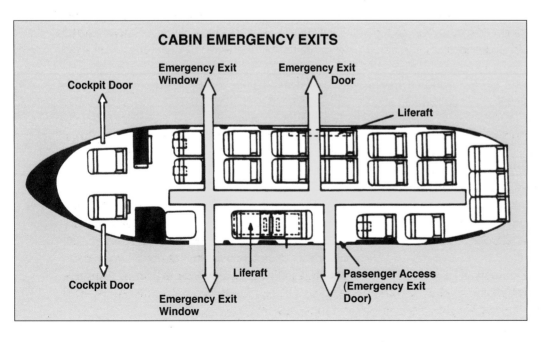

CABIN EMERGENCY EXITS

Cockpit Door

Emergency Exit Window

Emergency Exit Door

Liferaft

Cockpit Door

Liferaft

Emergency Exit Window

Passenger Access (Emergency Exit Door)

Figure 3.4 Typical seating and emergency exit arrangements on a SUPER PUMA/TIGER AS.332L helicopter.

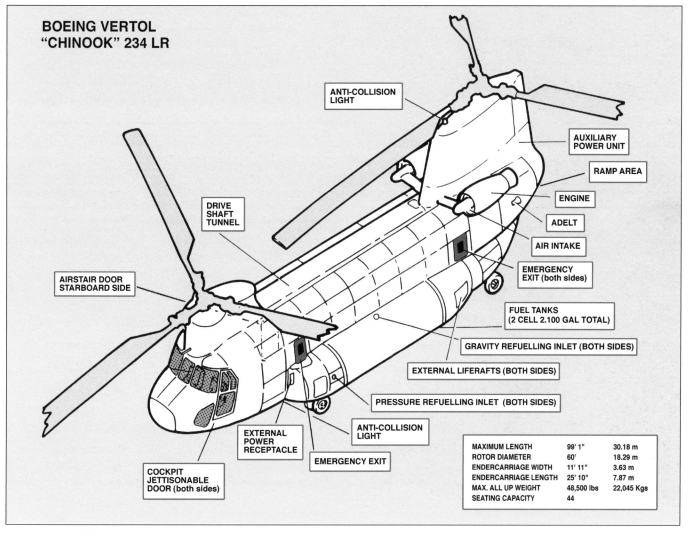

Figure 3.5 The various facilities for a Chinook helicopter. The military version may vary from this arrangement.

provided adequate precautions are taken against the ignition of fuel vapours, it should not prove difficult to cut-in, either through the metal struts or perspex area.

3.7 Flight Recorders

Whilst most helicopters are not usually equipped with flight data recorders, those operating in the North Sea carry flight deck and cockpit voice recorders similar to those fitted in fixed wing aircraft.

3.8 Military Helicopters

There are many different types of military helicopter in use today. Firefighters should be very wary when approaching these as they may be car-rying weapons. Most weapons will fire forwards, but the helicopter could be carrying pods on the side of the aircraft and these should be avoided if possible.

3.9 Special hazards

The hazards of weapons carried by military helicopters is dealt with in Chapter 8, "Special Hazards".

This section identifies some fundamental hazards associated with helicopter crash situations and gives advice on the dangers associated with such incidents. It cannot encompass every eventuality but should give firefighters an insight into precautionary safety measures and allied tactical awareness. These include:

The Fuselage Starts To Roll

Rotors Shatter

Passengers, fuel and exits become inverted with possible large fuel spillage

Figure 3.6 Typical effects of a helicopter crash.

- The deployment of handlines around a helicopter is far easier given its relative size. The application of media is unhindered beneath the airframe.

- In any crash situation it is unlikely that a helicopter will remain upright.

- There is every possibility that in a helicopter crash fuel will be released given the proximity of fuel tanks and lines in construction of the airframe.

- Given the relatively light construction of a helicopter it is unlikely to remain intact if it is involved in a crash.

- The construction of a helicopter includes a relatively high proportion of composite materials. Firefighters should ensure they have adequate respiratory protection when engaged in firefighting, and subsequently. (See Figure 3.2 and Chapter 8, Section 8.4.)

- The versatility of a helicopter can cause it to become involved in an emergency or accident virtually anywhere.

- Passengers, fuel and exits may become inverted with possible large fuel spillage.

Another danger associated with helicopters that are used largely over water such as the North Sea area is Water Actuated Devices (Buoyancy Bags). These bags are found in the wheel hubs or on sponsons which are braced to the aircraft fuselage by fixed struts or can be mounted on skids or can be integral.

They are intended to give stability to the helicopter when on the water, in the event that the aircraft has to ditch. These devices normally operate when immersed in salt water but may inadvertently be actuated by the application of water or foam during firefighting operations. (See Figure 3.7.)

ADELT

This is an Automatically Deployable Emergency Locator Transmitter which can be inadvertently actuated by firefighters during firefighting operations. It may cause injury as it is ejected away from the aircraft as it is deployed.

This unit is usually attached to the tail section on the starboard side and, when actuated, is deployed by a release mechanism downward and outwards, to the rear of the aircraft.

Summary

The design features of helicopters can result in early penetration of the passenger compartment by fire.

There is every likelihood that immediate and robust rescue and firefighting procedures will need to be implemented on arrival.

Figure 3.7 Water actuated devices (bouyancy bags) fitted in wheel hubs or sponsons. May be inadvertently activated by the application of water or foam during firefighting operations.

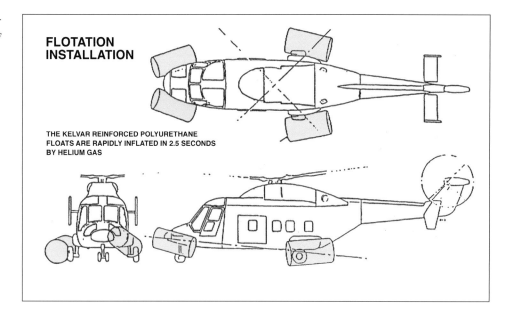

FLOTATION INSTALLATION

THE KELVAR REINFORCED POLYURETHANE FLOATS ARE RAPIDLY INFLATED IN 2.5 SECONDS BY HELIUM GAS

The construction of helicopters could result in passengers becoming trapped between the aircraft engines, gearboxes and fuel tanks. This can effect passenger survivability.

Some important factors that may improve survivability are:

● The training and experience of Rescue and Firefighting personnel.

● The effectiveness of the extinguishing agents and equipment, together with the tactics and techniques employed.

● The speed with which Rescue and Firefighting personnel respond and gain control over any resulting fire and effect rescue.

SAFETY

Firefighters should be aware that helicopters in crashed situations pose serious dangers to both themselves and to the passengers and crew who operate them.

Firefighters and personnel who may have to move around helicopters must be made aware of the dangers that helicopters present. Here are some of the more obvious:

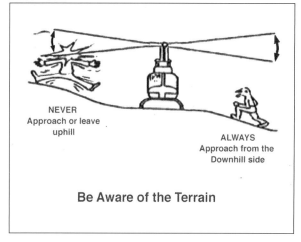

NEVER
Approach or leave uphill

ALWAYS
Approach from the Downhill side

Be Aware of the Terrain

Figure 3.8 Be aware of the terrain

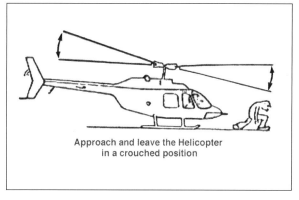

Approach and leave the Helicopter in a crouched position

Figure 3.9 Approach to and from the helicopter should be done in a crouching position.

Figure 3.10 The landing site for helicopters should be kept free from obstacles and loose materials.

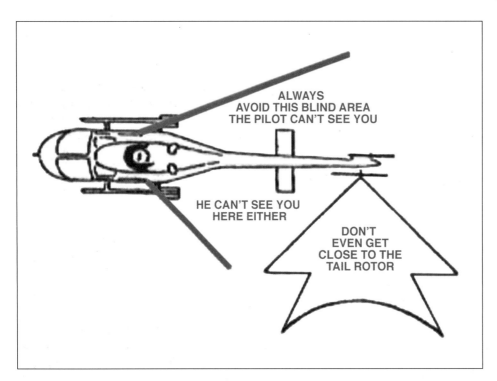

Figure 3.11 Avoid if possible these danger areas, as the pilot cannot see you. Beware the tail rotor blades.

Chapter 4 – Legislative Framework, Airports and Emergency Procedures

4.1 Legislative Framework

The International Civil Aviation Organisation was founded at the Chicago convention in 1944. The convention established international standards for the safe, orderly and efficient operation of global air transport. These standards and recommended practices are promulgated in a series of annexes encompassing a range of requirements for aviation safety. Annex 14 titled "Aerodromes" specifies the requirements for rescue and firefighting services at airports. In order to ensure these generic requirements are applied within countries, each has established its own national aviation authority. Within the United Kingdom this is the Civil Aviation Authority which is fundamentally governed by the Department of Transport. The Civil Aviation Authority has a statutory duty imposed by the Secretary of State for transport to enforce safety standards of economic, technical and operational performance in the aviation industry. Additionally, it provides navigational services at a number of airports and in conjunction with the Ministry of Defence, is responsible for the network of air traffic control, the National Air Traffic Service, NATS.

Regulation stipulates that aircraft flying in the United Kingdom for specified purposes, principally the public transport of passengers, may only use licensed, government or authority airports. A condition imposed on airport licensees is the provision of rescue and firefighting services in the event of an emergency at an airport or its immediate confines.

All civil airports within the United Kingdom are in private ownership. The British Airports Authority is the largest operator, maintaining seven airports, including Heathrow, Gatwick, Stansted and Southampton in England, and Edinburgh, Glasgow and Aberdeen in Scotland. Others such as Manchester, Birmingham, Liverpool and Luton are jointly operated by local authorities and private enterprise. Military airfields are considered in Section 6.

● **The Civil Aviation Authority enforces safety standards at civil airports in the United Kingdom.**

4.2 Categorisation of Airports

All civil airports are placed within categories ranging from Category 1, the lowest, to Category 10, the highest. Categories are fundamentally determined by the size of aircraft and scale of operations. The larger the aircraft and the greater the frequency of its movements, the higher the category of the airport. Correspondingly with the rise in airport category rescue and firefighting provision is increased. (See Tables 4.1 and 4.2.)

Airports such as Heathrow and Manchester operate at the highest category, whilst Southend and Biggin Hill are at a much lower level. Airport categories can change, however, as the scale of aircraft operation fluctuates. Brigades with airports on their ground should ensure they are familiar with the category of the airport, the extent of operations and most importantly the scale of rescue and firefighting provision maintained. Maps of the airport and the immediate vicinity with topographical details such as emergency water supplies, hydrants, access roads and break in gates should be available to responding brigades. Crews should familiarise themselves with the airport and any particular problems such as water hazards, rail transit systems or fuel storage depots. Liaison between Local Authority and Airport Fire Services in examined in Chapter 9.

Table 4.1

Aerodrome Category	Aircraft Overall Length	Maximum Fuselage Width
1	up to but not including 9m	2m
2	9m up to but not including 12m	2m
3	12m up to but not including 18m	3m
4	18m up to but not including 24m	4m
5	24m up to but not including 28m	4m
6	28m up to but not including 39m	5m
7	39m up to but not including 49m	5m
8	49m up to but not including 61m	7m
9	61m up to but not including 76m	7m
10	76m up to but not including 90m	8m

Table 4.2

Aerodrome Category	Foam meeting performance Level B			Complementary Agents		
	Water	Foam Concentrate	Discharge rate foam solution/minute	Dry powder	Halon	CO_2
	(L)	(L)	(L)	(kg)	(kg)	(kg)
1	230	14	230	45	45	90
2	670	40	550	90	90	180
3	1 200	72	900	135	135	270
4	2 400	144	1 800	135	135	270
5	5 400	324	3 000	180	180	360
6	7 900	474	4 000	225	225	450
7	12 100	726	5 300	225	225	450
8	18 200	1092	7 200	450	450	900
9	24 300	1458	9 000	450	450	900
10	32 300	1938	11 200	450	450	900

Airports are categorised by the size and scale of aircraft operations. The higher the category the more extensive the rescue and firefighting resources available.

4.3 Air Traffic Control (ATC)

All movements of civil aircraft to and from the UK and over it are regulated by the National Air Traffic Service (NATS). A branch of the RAF performs a similar function for military aircraft, but within the NATS control zones or areas such aircraft come under the additional control of NATS.

4.4 Emergency Procedures

4.4.1 Types of emergency

ATC classify emergencies under the following headings:

(a) Aircraft accident
An aircraft accident which has occurred on an airport.

(b) Aircraft accident off airport
An aircraft accident which has occurred off an airport but within two miles of its boundary.

(c) Aircraft accident imminent
An aircraft accident which has not yet occurred but is considered to be inevitable on, or in the vicinity of, the airport.

(d) Aircraft ground incident
An aircraft on the ground which is, or is suspected to be, involved in an incident which might endanger the safety of the aircraft or its passengers.

(e) Full emergency
An aircraft in flight which is known, or suspected to be, in such difficulty that there is a danger of an accident on landing. This category may also be used to classify unlawful acts, e.g., hi-jack.

(f) Local stand-by
An aircraft in flight that has developed some problem which is not considered sufficiently serious by the captain to prevent the aircraft making a safe landing, e.g., a hydraulic failure which may call for the utilisation of secondary or standby systems.

(g) Domestic fires and special services
A domestic fire is one within the airport boundary which might be dangerous to life, property, or aircraft operations. A special service call is when fire service personnel and equipment are needed to deal with incidents other than fires. (This category is not used when the safety of aircraft is involved).

(h) Aircraft unlawful act (Hi-jack)
An aircraft which may or may not be in flight and has been unlawfully seized.

(i) Aircraft bomb warnings
When information is received that a bomb is on board an aircraft which may or may not be in flight. The appropriate category, i.e., local stand-by or ground incident, will apply.

(j) Act of aggression – ground
An armed attack, bomb attack or suspected attack; the finding of suspicious objects or the taking of hostages on the airport which does not directly involve aircraft or their operation.

4.4.2 Preplanning procedures

One specific condition of the licence of a civil airport is a requirement to formulate Emergency Plans. Such plans are fundamental in achieving a successful outcome at the scene of a major aircraft accident. They must ensure adequate arrangements for the efficient alerting and mobilisation of all emergency services should an accident occur. All Brigades will have plans for airports and airfields in their areas and they will, of course, vary according to the size and complexity of the risk. At major airports it is usual to have several 'rendezvous points' to which the fire brigade, police, ambulance and other assisting municipal services will report to stand-by. For example, the London Fire Brigade plan for Heathrow includes four of these points strategically placed outside the airport boundary but close enough to access gates to ensure a rapid back-up to the airport fire services if required.

Various forms of communication system are set up at the rendezvous point(s) which enables the resources to be ordered on quickly. It is usual for an airport guide vehicle to lead appliances to the incident site when they are deployed 'airside', i.e. into the area where aircraft could be operating. This vehicle is in constant radio contact with Air Traffic Control in the control tower and will obtain all the necessary clearances required by vehicles moving airside. At smaller airfields, similar but simpler arrangements are made to direct appliances to the scene of an accident.

It is obvious that, in order to keep these plans and procedures contemporary, they need to be practised on a regular basis and up-dated as necessary. Brigades should ensure that all personnel likely to attend are full conversant with the plans.

Emergency planning is fundamental in achieving a successful outcome at the scene of a major aircraft accident.

4.4.3 Calls to incidents (Basic Procedures)

The ATC officer on duty is responsible for alerting the emergency services in the event of an accident or other emergency. This is usually done by a telephone link to the airport fire service control room if the incident takes place on or near the airport; the call will then, if necessary, be relayed to the LAFB control room by the airport fire service control room attendant. If the incident occurs away from an airport, ATC will notify the police, who will in turn inform the fire brigade.

Sometimes, brigades may receive calls from other sources. If so, they should notify the police immediately.

4.4.4 Attendances

Generally at aircraft incidents on airports or airfields brigades will augment the airport fire service. Although the LAFB have the statutory responsibility under the Fire Services Act 1947, it is usual for the senior airport fire service officer to be in command initially. When a senior LAFB officer arrives at the incident he will assume command of the combined operation. However, he should not alter the deployment of airport appliances, equipment or manpower without very good reason, as the airport fire service have tactical plans which they put into operation to cover certain eventualities.

At incidents occurring off an airport but within two miles of one, the airport fire service may attend in a reduced capacity, depending on the distance from the airport. In these circumstances, the LAFB will normally take full tactical control of the incident on their arrival. At distances of more than 3.3 kilometres, airport fire services do not usually attend unless requested to do so by the LAFB. Such contingencies should be considered with emergency plans, where a call on airport firefighting resources may prove invaluable at a major aircraft incident.

Where an airport is located near large areas of water, airport fire services are required to provide water borne rescue equipment. (See Chapter 6, Section 6.7.2.) The amount of equipment provided will vary in line with the scale of operations at the airport but it is generally limited. The Emergency Plan will provide for the alerting and mobilisation of assistance from marine organisations such as the RNLI, RYA and The Coastguard. The Armed Services may also be called upon to help in providing air sea rescue helicopters.

4.5 Civil Airport Firefighting Facilities

4.5.1 Appliances

The appliances attached to airports and airfields will vary according to their media carrying capacity and related discharge rates, generally governed by the airport category. However they must all meet with International Civil Aviation Organisation (ICAO) specifications. These are laid down in Civil Aviation Publications (CAP 168 for the UK). Minimum requirements with regard to performance, acceleration and braking, cross-country capability and, consequently, ground clearance limits, are all specified.

Large capacity water and foam tanks are the norm, giving the larger appliances a huge firefighting capability without recourse to outside supplies. An important feature of the firefighting potential of an airport fire appliance is a design requirement that the amount of foam concentrate carried must be sufficient to supply at least two full loads from its water tank. This dimension is of primary importance in establishing operational tactics at the scene of an aircraft accident/incident. **The rapid replenishment of airport fire appliances with water by responding local authority appliances will allow foam production to be sustained and/or the security of the fireground to be safe-**

guarded. Complementary firefighting agents such as dry chemical and halogenated hydrocarbon may also be carried in bulk supply. A gross weight of 35 tonne is not unusual.

Attendance times – as little as two minutes to any part of an airport, with a given firefighting capacity – make the need for rapid acceleration obvious. High discharge rates of finished foam from appliance monitors to a distance of 90 metres, a capability to produce foam whilst moving at slow speeds, all wheel drive capability and the provision of hand lines are all standard requirements.

One development is the introduction of light foam tenders. These may be termed *"First Strike"* appliances, being utilised to respond rapidly to incidents and initiate fire suppression and rescue operations whilst awaiting the arrival of major foam tenders. Light Foam Tenders have rapid acceleration and high speed capabilities coupled with considerable manoeuvrability. They are usually equipped with monitors having the same range and discharge capabilities of their larger counterparts, whilst their water carrying capacity may be in the order of 6000 litres. (See Figure 4.1.)

The amount of foam concentrate carried by an airport fire appliance is sufficient to supply two full loads from its water tank.

4.5.2　Water supplies

Water supply systems at airports and airfields vary considerably but are usually classified under two headings: (i) primary and (ii) secondary.

(i)　The primary supply consists of a below-ground water mains system which is usually fed from street mains of the local water authority. The size of the airport main is therefore generally determined by the size of the water authority's supply, usually 150 mm or larger. Often the system is a ring main which reduces friction loss, allows water to flow in either direction, and enables different sections to be isolated without interfering with the supply to others.

(ii)　Secondary supplies may be provided, in the form of high-level or underground tanks or sub-surface reservoirs. These are often fed by storm water catchments or piped supply.

Figure 4.1 Light Foam Tender (First Strike Appliance). Water carrying capacity 6000/7000 litres with monitor output comparable with that of a Major Foam Tender.

Figure 4.2 A Typical Major Foam Tender with water capacity in the order of 15,000 litres and 2,000 litres of foam compound.

Ponds, rivers and streams may also be used as supplementary sources of water.

Fire hydrants are placed at strategic points around airports, their layout depending on the risk. They are usually indicated by hydrant plates, and, although most of the outlets are below ground level and require standpipes, some airports have provided pillar hydrants. All this information should be indicated on the airport grid maps.

Water supplies to airports may be metered, and it could be necessary to operate a valve to by-pass the meter. Although airport fire service personnel will know of these valves, LAFB firefighters should also make themselves familiar with their location and operation.

Firefighters should familiarise themselves with airport water supplies.

4.5.3 Hangar protection

(a) General

Whilst major airports are equipped with efficient firefighting facilities for aircraft landing, taking off or taxying, they also have to cover the aircraft maintenance areas. An aircraft on a fairly short inspection, or period of maintenance, may be installed in a hangar with a full fuel load. The failure of a component during maintenance could result in a major fuel release in a confined space. Apart from this, the presence of highly volatile solvents, large amounts of electrical apparatus and a big human element all require a high degree of protection, not only for the occupants and the building but also for the aircraft. It is therefore important for hangars to have adequate equipment for suppressing any incipient fire. The National Fire Protection Association's Standard 409, "Aircraft Hangars", is closely followed in the UK, and parameters relating to hangar size, aircraft size and even aircraft design are used to categorise hangars for fire protection. Firefighters should visit the hangars at local airports and familiarise themselves with their layout.

Fire protection equipment for aircraft hangars falls into two basic categories: high-level and low-level.

(b) High-level protection

For large risks, there are three main systems used:

(i) High expansion foam discharged to cover the entire floor area to a depth capable of enveloping the aircraft.

(ii) Low expansion foam discharged for a given period by means of a deluge system mounted at hangar roof level.

(iii) Discharge of foam/water solution using AFFF through the standard automatic sprinkler system.

In system (ii), the roof is divided into zones, delineated either by the type of construction or by draught curtains. Each zone contains a number of specially designed sprinkler heads, and a deluge control valve which enables it to operate as an independent unit. The number of zones brought into operation depends on the extent of heat spread from the fire.

(c) Low-level protection

When a high-level deluge system operates, aircraft, by their design shape, cause an 'umbrella' effect in that any fire under the wings or fuselage cannot be directly affected by the extinguishing medium. Any aircraft with a wing in excess of 280 sq metres requires supplementary protection. A Boeing 747, for example, has a wing area of 510 sq metres, and Concorde approximately 360 sq metres.

The most satisfactory method of providing protection under the wings and fuselage is by either fixed or oscillating foam projectors. (See Figure 4.3.)

Sometimes hangars have special maintenance platforms which follow the shape of the leading edge of the wings of individual aircraft. (Figure 4.4.) Here, foam projectors are fixed onto the frame of the platform, cutting down the distance of throw and providing a faster attack on the fire. In other hangars, where aircraft of various types are maintained, the foam projectors may be mounted on the walls of the hangar.

Besides foam, there may be halon, dry chemical or CO_2 installations for particular risks. Also, to deal with any small fires, there are usually portable extinguishers placed at strategic points.

(d) Detection

There is a sophisticated system of fire detection equipment in all large hangars. Various types will be found. Some hangars use a system which responds to an abnormal rate of rise in temperature, and which is unaffected by stray ultra-violet or infra-red rays. Usually, total reliance on one system is not recommended, and this system is

Figure 4.3 Example of underwing and overall protection from fire in a hangar.

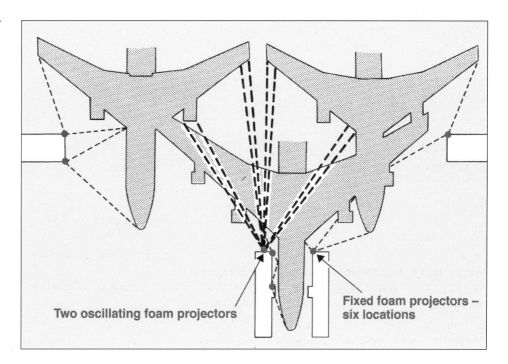

Two oscillating foam projectors

Fixed foam projectors – six locations

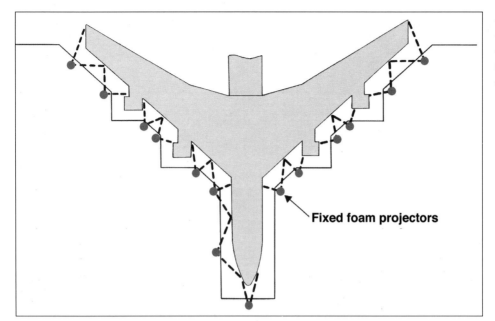

*Figure 4.4
Fire protection in
hangars where fixed
maintenance platforms
are used.*

Fixed foam projectors

often backed up by another using quartzoid bulb detectors with a fixed operating temperature.

4.6   Military Airfields

4.6.1   Categorisation and firefighting facilities

Military Airfields are divided into five fixed-wing and three rotary wing categories. Generally the larger the size of aircraft/helicopter the higher the category of airfield and consequently the rise in rescue and firefighting provision. Whilst the categorisation of Military Airfields follows a similar pattern to that used for Civil Airports, specific factors are additionally considered; these include:

(i)   Passengers
(ii)   Armaments, including ejection seats
(iii) Dangerous air cargo

Firefighters stationed on these airfields are specially trained in crash fire rescue procedures, having particular expertise in dealing with military aircraft and their associated risks. The organisation, equipment, manning and training is geared to a response time of two minutes to an aircraft crash and a control time of one minute from the start of firefighting operations.

Appliances vary but, on the larger, busier airfields, are similar to those used on a large civil airport.

Military fire services are also used to cover the domestic risks on their airfields and possibly, in the adjacent areas which may be military property, e.g., living quarters. The attendances for these risks will be as for any other similar risk category.

Airfields used by the USAF have similar arrangements for fire cover. Their categorisation is comparable with UK Military airfields, being set by the United States Department of Defense. Five categories or "indexes" are determined; being A to E. Again with the ascent of the index of the airfield rescue and fire fighting provision is increased. The D.O.D additionally requires that U.S. military installations are equipped to deal with structural and hazmat emergencies. Rescue and fire fighting resources at such airfields will therefore be extensive with some fire appliances weighing in the order of 60 tonnes. Notwithstanding such capacity, the LAFB will form an integral part of a military airfield's emergency plan.

4.6.2   Calls to incidents

A call to a brigade about a military aircraft accident will usually be made by the RAF, RN or

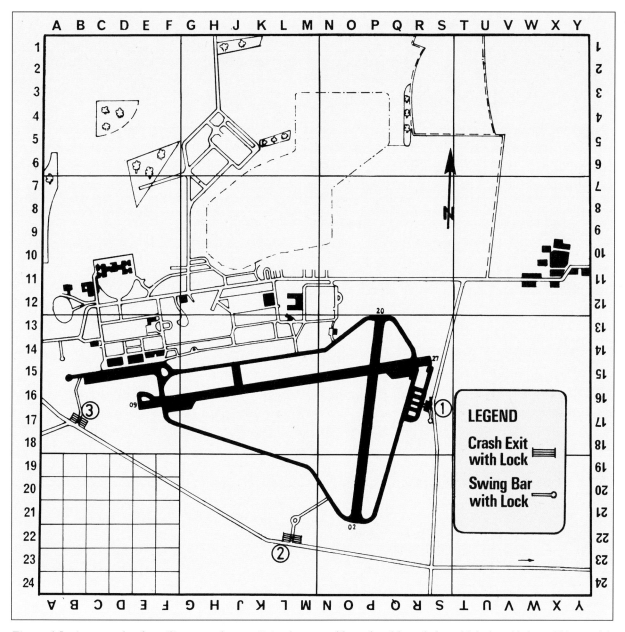

Figure 4.5   An example of a military crash map. Brigades can add any local knowledge which they think could be useful.

USAF via the police. If a brigade receives a call from a different source, they should notify the police immediately, asking them to inform the appropriate military authorities and to obtain from them details of any special hazards involved. (See Chapter 8, Section 8.3.1.)

4.6.3   Liaison with military authorities

In each brigade with a Military or USAF airfield in its area, the officer-in-charge should set up the same sort of liaison with the local station commander as with the manager of a local civil airport.

He should obtain, if possible, a copy of the airfield crash map (Figure 4.5). An effective plan of co-operation should be drawn up and responsibilities allocated. There will be areas, of course, which are militarily sensitive, including some aircraft, but these will be matters to be negotiated at whatever level is appropriate. The military authorities, as well as local authorities, will have to bear in mind that, in the event of a large incident, e.g. weapon storage depot fire, the resources of the brigade will have to be employed. Without knowledge of the risk, firefighters will be at a disadvantage, especially regarding personal safety.

Aircraft Incidents

Chapter 5 – Incidents on Airports

5.1 General

The primary object at an aircraft incident is to save life, and to do this by creating conditions where survival is possible. The extinguishment of a fire may continue after people have been rescued or given time to evacuate the aircraft safely. This chapter deals with firefighting procedures and operational tactics and techniques designed to assist survival at aircraft fires occurring on airports; actual rescue procedures are dealt with in Chapter 7.

The great speed at which fire can develop at an aircraft incident has led to the evolvement of the appliances, equipment and techniques which the airport fire services constantly use in training and update. The rapid back-up by the local authority fire brigade is vital in many incidents and at any major accident the resources of the airport fire service will be extended, particularly in terms of personnel. There will generally be a need for rapid replenishment of crash fire rescue vehicles in order to sustain firefighting operations and/or to ensure post fire security. (See Chapter 4, Section 4.5.1, Appliances).

Aircraft fires are not always external and do not always involve fuel, although the risk of the fire spreading to the fuel can never be ruled out. An internal fire in flight or a similar emergency situation creates an urgent need for the pilot to land his aircraft at the nearest airport affording adequate facilities for the particular aircraft type (runway length). This usually gives the airport fire services time to deploy, alert local authority fire brigade and other agencies, and creates the best conditions for rescue and firefighting. Unfortunately many accidents happen either on landing or take-off, and the response, although fast, often approaches a rapidly deteriorating situation. (Sometimes, of course, where a mishap occurs in flight, the pilot is unable to reach an airport in time, and the aircraft subsequently crashes or lands elsewhere. This eventuality is dealt with in Chapter 6).

5.2 Features of Aircraft Fires

An aircraft taking off will have a large fuel load, and even at the end of a flight there is still a considerable amount of fuel on board. Any serious crash will inevitably lead to an escape of fuel which will often ignite immediately. There are, moreover, numerous sources of potential ignition in a damaged aircraft which are capable of starting a fire some time after impact. The development from ignition to peak intensity can be very rapid, the fuel flowing and presenting a large flame front and huge amounts of smoke. (See Figure 5.1.) The aircraft is quite likely to be on its belly, possibly in several broken sections, perhaps on grass or off the end of the runway.

A considerable portion of an aircraft is occupied by highly flammable fuel systems. Any escape of fuel may ignite immediately.

5.3 Basic Firefighting

5.3.1 Positioning of appliances

The first fire appliances to arrive will invariably be those of the airport fire service. The officer-in-charge must deploy his appliances and crew immediately to the best advantage, taking particular care to keep the aircraft's doors and escape slides usable for passengers to escape. The ability of appliances to apply foam or other extinguishing media whilst on the move means that a developing, moving situation can be constantly covered. Wind may keep one side of the aircraft relatively clear, but all parts of the fuselage are vulnerable to

Figure 5.1 Aircraft fires can develop rapidly with a large flame front and huge amounts of smoke.

penetration by fire, and heat can even cause internal fittings to decompose and burn, giving off toxic fumes. Figure 5.2 gives examples of appliance positioning, emphasising the primary objective of protecting the integrity of the aircraft's fuselage and keeping fire from the occupied cabin to allow evacuation. The desirability of positioning appliances upwind is also clearly illustrated.

The aircraft may have come to rest off the runway and on very soft terrain where it would be difficult to manoeuvre appliances. In these circumstances, it may be necessary to park the appliances on the nearest solid ground and deploy hand lines. **Firefighters will need to bear in mind however, that free fuel may be flowing around them, and, if not adequately covered by foam or another extinguishant, could ignite and trap them.**

Similar care should be taken in respect of appliances. If any of the aircraft's engines are still running, all personnel should keep well clear of their intakes and exhaust outlets (see Chapter 7, Section 7.2.1(a) and Figure 5 3).

5.3.2   Application of extinguishing media

The main objective of firefighting will be to keep the fire away from the escaping occupants of the aircraft without obscuring or hampering any escape route. The escape slides of the aircraft will have been activated and passengers will be using them (Figure 1.13). Some passengers may make their way along the wings or down the wing roots, and any application of foam or water onto these areas could impede their escape and make the surface dangerously slippery.

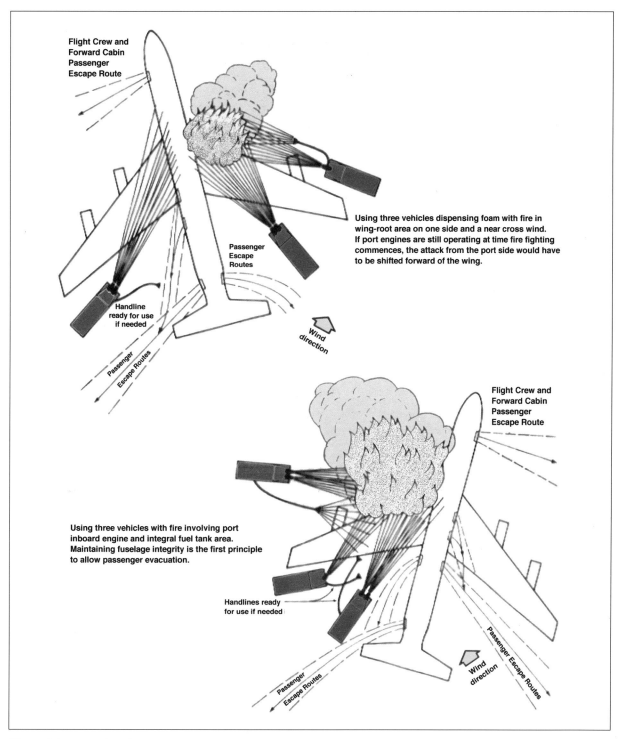

Figure 5.2 Principles of aircraft firefighting, highlighting appliance positioning and the application of foam aimed at protecting the integrity of the fuselage.

Sufficient foam should be applied to the fire to knock down the flames and maintain a seal preventing re-ignition. Care should be taken not to drive fuel about by too direct or forceful an application. If fire penetrates the fuselage, decisive firefighting action should be taken to introduce a water fog into the affected area, both to cool it and to prevent the rapid build-up of heat and smoke which can trap people inside. It may be necessary to penetrate a window in order to insert the fog nozzle.

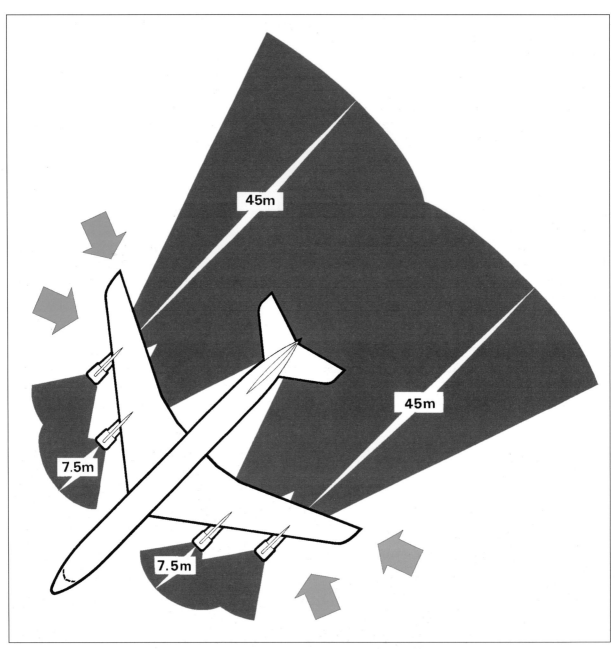

Figure 5.3   Personnel should keep well clear of engine intakes and exhausts.

Elevated waterways or aerial nozzles are being introduced within the United States following extensive evaluation and testing by the Federal Aviation Administration. This equipment is not only capable of applying water spray within the cabin interior, it normally incorporates a penetrating lance capable of piercing an aircraft's fuselage. This allows for the rapid application of media without hindering evacuation allowing firefighters to subsequently enter with handlines under more tolerable conditions. Such equipment is currently being evaluated in the UK and Europe. (Figure 5.4.)

If the fuselage does become smoke-logged, ventilation will be necessary. The requirement to enter for this purpose, and yet not impede passengers escaping, can call for a quick decision as to the method and point of entry.

Firefighters must bear in mind, however, that to ventilate without a cooling fog may lead to a backdraught.

There could be areas where foam will not penetrate, and there is a requirement for suitable extinguishants to consolidate the initial attack. These may be carbon dioxide, dry powder or a halogenated agent.

Not all aircraft crashes are of sufficient severity to result in fire, but any free fuel is a potential danger and should be covered with foam to prevent ignition. If fuel is leaking from ruptured tanks or supply lines, firefighters should attempt to seal off the flow. In any case, sufficient equipment should be deployed and manned so that, if a fire does break out, it can be suppressed quickly.

Firefighting tactics utilising wind, terrain, appliance positioning and the correct application of foam are aimed at protecting the integrity of an aircraft fuselage.

5.3.3 Handling of aircraft equipment

Airport firefighters should have the knowledge to be able to isolate master switches, disconnect batteries, and secure an aircraft undercarriage and generally make equipment safe. Local authority firefighters attending airport incidents should not operate any of this equipment.

5.4 Particular types of fires

5.4.1 Metal fires

Given the usual attendance times at an aircraft incident on an airport, the likelihood of a fire having reached a point where the metal is actually burning is remote.

Aluminium will ignite at around 800°C, but the various alloys, e.g. duralumin, do not usually ignite in aircraft fires, possibly because their melting points are well below their ignition points and the metal therefore has time to flow away from the heat source.

Figure 5.4 An elevated waterway incorporating a penetrating lance. This equipment allows for the rapid application of water spray within the cabin interior without hindering evacuation.

Magnesium alloy is usually in large sections, in which form it is not easily ignited. If it does ignite however, it can burn violently, and any initial application of foam will cause it to burn explosively. Nonetheless, a determined attack with a large jet of water will quickly accelerate the release of hydrogen beyond a limit where combustion can be sustained, with eventual cooling taking place. The reaction with foam is liable to be more violent with extinguishment being prolonged.

Titanium and stainless steel, in order to ignite, need a sustained temperature in excess of 2000°C, which is unlikely to be reached in the course of firefighting.

5.4.2   Brake and wheel fires

The overheating of brake and wheel assemblies usually follows excessive braking. This can happen on landing when a hydraulic failure reduces the ability of the pilot to utilise control surfaces (flaps) and he feels the aircraft may over run the runway or more often, on an aborted take-off. Quite often tyres will burst or, as the aircraft comes to a standstill, ignite by transference of heat from the brakes. To avoid metal failure, firefighters attending overheated brake and wheel assemblies should usually refrain from applying water and allow natural cooling. If there is fire however, a fine water fog applied from fore and aft of the undercarriage assembly is generally used, **personnel and appliances being kept away from the side of the wheels in case of rim failure, which may take place with explosive force**. If there is any possibility of undercarriage failure, the officer in charge should ensure locking pins are inserted. These should be available from the airline crew or ground engineers. Personnel working beneath an aircraft should always be kept to a minimum particularly when an undercarriage has sustained any form of damage.

Personnel working beneath an aircraft should be kept to a minimum.

5.4.3   Ground incidents

Occasionally, stationary aircraft or others taxying to and from runways are involved in incidents, some involving fire. These may occur owing to a fuel spillage during refuelling or collision with another aircraft, vehicle or airport structure. There may be passengers on board the aircraft and whether attached to a loading jetty or not escape slides may be deployed. Drivers should be cautious of driving through smoke plumes as evacuation may be taking place and they risk causing casualties. Horns, warning lamps and vehicle lights should be fully utilised and even when appliances are finally positioned, the lights should be kept on until the smoke has cleared to warn other responding vehicles approaching the incident.

Figure 5.5 Typical damage to undercarriage caused by an aborted take-off.

Chapter 6 – Incidents off Airports

6.1 General

When an aircraft in flight develops an emergency condition, the captain will make every effort to land at the nearest suitable airport. He may, however, be forced to make an emergency landing elsewhere, and this may prove particularly hazardous if there is a shortage of flat, open land in the vicinity. Occasionally, a structural failure may cause an aircraft to plunge suddenly and uncontrollably to the ground, possibly in a built-up area. The latter type of accident can obviously result in much loss of life, injury, and damage to property.

6.2 Liaison

The first responsibility for liaison with civil and military air authorities in respect of aircraft in distress or crashed in the United Kingdom (other than on airports) lies with the police. On receiving information from ATC about an aircraft in distress or which has crashed away from an airport, the police immediately notify the other emergency services, giving all relevant information.

If a brigade receives such information from a different source, they should notify the police as firefighting resources are mobilised; the police will then notify the appropriate air authorities, the ambulance services, and any other authorities they think necessary. The information should contain as much of the following as can be ascertained:

(i)   Type, description, registration number and flight number of the aircraft;

(ii)   Number of passengers and crew on board;

(iii)   Nature of the incident;

(iv)   Location of incident;

(v)   Time incident occurred;

(vi)   Action already taken;

(vii)   Name and location of officer making report;

(viii) Name and address of person making report.

Brigades should ensure the police are alerted to an aircraft accident away from an airport should they receive such information from a different source.

6.3 Locating the Incident

The speed with which a crashed aircraft can be located will depend on a number of factors. If the captain has been able to give the position to air traffic control or if the aircraft has been seen to disappear off the radar screen at a certain point, the circumscribed area could be relatively small. Unfortunately some accidents happen very quickly, leaving the crew little or no time to send a distress call, and this often results in a call to the emergency services from an observer who may only be able to give a rough estimate of where the aircraft has crashed. Obviously, whoever takes the call must make every effort to get as much information as possible, especially the exact position of the observer. Most authorities, including the RAF airfield services, use the National Grid of the Ordnance Survey and, in any preplanning, a general agreement on map references based on this system is invaluable. The chances of the observer being able to give a grid reference are remote but, from the information gleaned, a search area based on a calculated reference point could be given to the PDA (pre-determined attendance).

6.4 Approaching the Incident

If the crash is in an inhabited area, location and approach could be relatively simple. It is when the crash has occurred in a remote rural area that local knowledge is particularly important. In deciding on the route of approach, a lot will depend on the terrain, time of year, weather etc, and it will probably be expedient to divide the PDA and try alternative routes if there is any doubt about a particular way in. Most brigade appliances are not designed for rough cross-country work and care must be taken not to get them bogged-down, blocking the approach to an incident. Regard should also be made of their ground clearance. Vehicles left unattended whilst personnel proceed on foot should be parked clear of the approach in a safe manner.

There will be a considerable amount of debris to be avoided, particularly if the approach follows a slide path made by the aircraft. In all cases, a sharp lookout must be maintained for survivors who may have been thrown from the aircraft or crawled away.

Once the incident has been found, Brigade Control should be informed of the exact location, the best approach and the operational situation e.g., fire, scattered wreckage, the presence of any survivors, the assembly points for appliances and ambulances etc., and the need for any further assistance. If possible, a guide should be posted or, at night, portable lights positioned, to indicate the route in and, perhaps, an alternative route out.

> Once the incident has been found, Brigade Control should be informed of the exact location, the need for further assistance, the best approach and the operational situation.

6.5 Firefighting

At a crash in an inhabited area a lot will depend on where and how the aircraft has finally come to rest but, generally, the firefighting will be similar to that described in the previous chapter, with the initial firefighting being the application of foam.

However, at incidents in remote rural areas involving heathland, woodland, crops etc., a fire may have been burning for some time before appliances can reach the scene, and there may be problems of water supply, wind direction and approach route. **The officer-in-charge of the initial attendance must carry out a dynamic risk assessment and may have to decide whether to:**

(i)   tackle the main bulk of the wreckage fire with the very limited resources at his/her disposal (Figure 6.1);

(ii)   wait for reinforcements whilst personnel look for survivors, and then set up an attack;

(iii)   use available water to prevent the spread of fire which could inhibit reinforcing appliances attending.

The officer-in-charge must, as previously stated, give Brigade Control all the information on the situation, state what assistance will be needed, direct the approach of appliances, establish rendezvous points, liaise with the other emergency services as appropriate, look for water supplies and consider setting up a water relay.

If possible, appliances should be positioned as for a similar accident on an airport. This will depend on the terrain, although the Incident Commander should endeavour to keep appliances on higher ground and upwind so as to avoid fuel spills and fuel vapours from ruptured tanks. The Incident Commander will also need to reconsider his tactics should the aircraft be military and possibly armed (see Chapter 8).

6.6 Casualties

6.6.1   Location of casualties

Even in the most severe of aircraft crashes there could be survivors, albeit badly injured, and a careful search should be made over a fairly wide area along the slide path and around the final wreckage point. In the time it could take the crews to find and reach the scene, people may have escaped from the aircraft, moved some distance away, and then collapsed, possibly in ditches, under bushes etc. Some people may be lying injured after being thrown out of the aircraft on impact. Many, of course, may still be inside the aircraft or under debris, being either physically

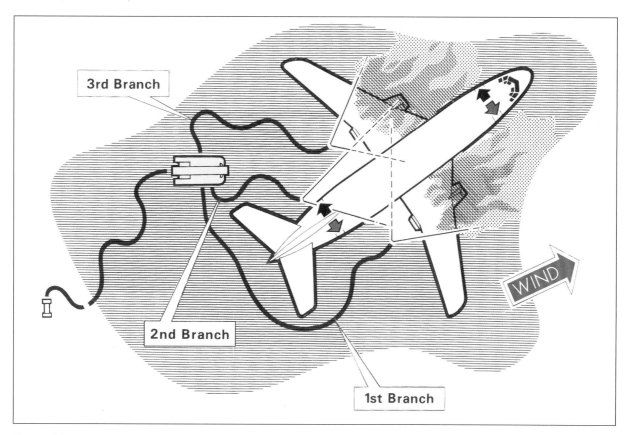

Figure 6.1 Tactical deployment with limited resources. One appliance using handlines to protect the integrity of the aircraft's fuselage.

trapped or in an unfit state to escape. Methods of rescue are dealt with in Chapter 7.

It should be possible to obtain, from ATC, an indication of the number of people who were aboard the aircraft. Firefighters should remember that, if ATC can be given the registration number of the aircraft or even the aircraft type or operating company (airline), this will help to identify the flight in cases where ATC do not already know it. If the number of people on the flight can be ascertained while the brigade is still on the scene, the incident commander should check this against the number of survivors and bodies located. It should, however, be remembered that not all the accident victims will necessarily have been occupants of the aircraft, since there may be ground casualties also, especially where a crash occurs in a built-up areas. The total number of people found should be carefully recorded, and this information should then be passed on to the police, who have the ultimate responsibility for ensuring that everyone is accounted for.

6.6.2   Handling of casualties

The handling of casualties in any off-airport incident will be difficult, but the advice given in Chapter 7 Section 4 should be followed. If the Ambulance Service are already in attendance, casualties should as a general rule be carried direct to waiting ambulances. If these have not yet arrived, then a safe area away from hazards and further possible involvement in fire should be set aside where survivors can, in the meantime, be carefully taken and given first aid.

Firefighters should however bear in mind that some casualties may have injuries which could be seriously aggravated by inexpert handling; it may be desirable, depending on the fire situation, to leave such casualties in situ until the arrival of the ambulance service. (This is particularly relevant in cases of ejected crew members from military aircraft, as there is a probability of sustained back injuries).

In any aircraft crash, almost all survivors are likely to be suffering from shock, and those that are able to walk will therefore need to be physically led to the ambulances or casualty clearance area, as they may not respond to directions.

It will be difficult, in a multi-casualty incident, for firefighters to decide which casualties to deal with first. Even identifying who is dead and who is alive will not be an easy task; casualties should be assumed to be alive unless it is clearly obvious that they are not. Firefighters should ensure that the other emergency services are informed of how each casualty has been dealt with. Questions relating to dealing with the deceased are dealt with in Section 6.8 – Post Accident Management.

It may be prudent, given the prevailing circumstances to leave seriously injured casualties in situ until the arrival of the Ambulance Service.

6.7   Particular Types of Accident

6.7.1   Crashes involving buildings

Where a passenger aircraft crashes into buildings, especially those which are inhabited, the problem normally assumes the classification of a **major incident**. (See Figure 6.2.) The breaking up of the aircraft could spread fire and destruction over a large area, with several major fires and many small ones developing at the same time. Apart from the aircraft casualties, the numbers involved in the buildings could be high, and the county major accident scheme should be implemented as soon as possible. (See the *Manual of Firemanship* Book 12, Chapter 4).

If the fuel has not ignited, it should be covered with foam and all persons in the vicinity and downwind for some distance warned to put out fires, and avoid smoking or any use of flame. It is quite likely that some of the fuel may enter drains, therefore the appropriate Environmental Agency must be informed of this risk, and the sewers flushed with water.

6.7.2   Crashes into water

Where airports are situated near large areas of water, there is always the possibility of an aircraft crashing into the water. Airport fire services which have identified this risk, will have contingency plans that include the use of water borne rescue equipment. Should airports not have such equipment and a potential water hazard prevails, arrangements will have to be made with the Coastguard, RNLI, local pilot organisations, tug companies, or any other marine agency that can render assistance. These arrangements will form part of the airport's contingency plans.

When aircraft crash into water, there is the possibility of extensive fuel leakage onto the surface of the water, which may or may not ignite. The breaking up of this fuel is important and can be achieved by using jets of water, if this is not practical or is unachievable then the fuel must be covered using foam, this will prevent ignition and make the spread of the fire less likely. Firefighters should remember that fuel, rising to the surface, may come into contact with hot parts of the aircraft and other ignition sources such as boat engines. It is likely to remain a potential hazard for the duration of the rescue operation, although it will disperse eventually.

Depending of how the aircraft came down, sections of the fuselage may be found floating with passengers still inside, and care must be taken to avoid disturbing their capability to float. Rescue operations should be carried out as quickly as possible in case buoyancy is lost. Rescuers should themselves beware of being trapped inside the fuselage. Occasionally, occupied sections of the aircraft may be submerged but still retain their integrity and, depending on the depth of water, divers may be required urgently. The incident commander should liase with the police on this matter.

6.8   Post-Accident Management

6.8.1   Introduction

Once rescue and firefighting operations have been completed, the responsibility for the security of the accident site, the wreckage, its contents, personnel

Figure 6.2 El Al Boeing 747 crashes into apartment block in Amsterdam, October 1992.

and other effects will be that of the Police Authority. The Air Accident Investigation Branch of the Department of Transport have the authority to investigate aircraft accidents and will be informed of any such incident by the Police. Their role is to determine the facts, conditions and circumstances leading to the accident with a view to establishing the probable cause.

Rescue and firefighting personnel should be fully aware of the scope of the investigation and enquiries undertaken, and that the accident may become a criminal investigation. All personnel at the scene should therefore conduct themselves with some thought of the subsequent investigation.

6.8.2 Dealing with Bodies and Recording of positions and locations

The position of fatally injured victims whether in the aircraft or not are extremely important for identification purposes and also to help establish the cause of an accident.

The removal of bodies should only be carried out under the direction of the medical authorities in conjunction with the Air Accident Investigation Branch; this is to ensure that body positions and locations are accurately charted.

The removal of the bodies prior to the arrival of the Air Accident Investigation Branch or medical teams may however be necessary in order to facilitate the rescue of survivors or to prevent the bodies being destroyed by fire or by some other hazard.

Where this is the case, the position of the body and its location should be noted, labelled, if possible, and reported to the investigation team. Rescuers who have moved bodies should be questioned and a statement should be made as soon as possible after the accident, whilst the memory of their actions is relatively fresh and they can recall body positions fairly accurately. Whenever possible, an officer should be appointed to map out as accurately as possible the location and position of bodies, although it must be borne in mind that some incidents, especially high speed crashes, produce carnage over a wide area with human remains being unrecognisable as such.

It should also be remembered that bodies that have been badly burnt become brittle and are likely to fall apart if untrained personnel move them. This can destroy vital evidence of identification and pathological evidence of the cause of death.

If cameras are available then photographs or video should be taken of the wreckage, the accident site and the position of the bodies. Any photographs that are taken by the press may also prove useful to the investigating team although the media should not be allowed to wander around the site without supervision.

6.8.3 Personal effects and Documentation

Most passengers on aircraft carry a number of personal effects and papers, such as passports, driving licences, reservations etc. The position of these documents and personal effects of both passengers and crew of the aircraft and any documents and papers that are scattered around the incident site may help the investigating team and it is therefore important to leave these articles in situ or where they need to be moved, their location and position should be recorded.

It is important to control the number of people allowed on the incident site, so that evidence such as personal effects are not disturbed, or are disturbed as little as possible. When the situation permits, there should be a careful withdrawal of all personnel and vehicles that are not essential.

Where bodies are moved, great care should be taken to ensure that any item, which is adjacent to the body, is recorded or moved with the body. Any items that falls from the body whilst being moved, should be collected, recorded and kept with that body as it may prove to be a means of identification.

6.8.4 The Movement of Wreckage

The wreckage of an aircraft must not be moved or disturbed until the Air Accident Investigation Branch gives its permission. If the aircraft, its wreckage or any part has to be moved before the investigation is complete, then a record must be made of the location and position of all the parts exercising specific care to preserve any evidence which may be crucial to the investigation e.g., position of controls, levers, switches, etc.

Once permission has been given to move the aircraft or wreckage, and the necessary safety measures embraced, the interior of the aircraft should be thoroughly ventilated.

Any surfaces that may have been contaminated with fuel or flammable liquid spills, including runways or ground surrounding the accident, should be flushed and once the aircraft has been removed, surfaces should be rechecked before normal traffic is resumed.

6.8.5 Defueling

Any fuel that remains in the aircraft's tanks should be drained prior to the removal of the wreckage or before the removal of fuel containers from damaged sections of the aircraft. Qualified engineers should drain any fuel that is removed, who should measure the fuel removed and retain the samples for analysis by the investigating authorities. When defueling is carried out, a firefighting vehicle and crew should stand by at the site.

6.8.6 Preservation and Gathering of Other Evidence

(a) Official Documents

A large number of documents are carried on aircraft, and rescue and firefighting personnel should carefully gather any documents that they find and hand them over to the Police or the investigating team. Some of the documents may be burnt, partially burnt or damaged in some way and they should be collected and placed in a plastic bag which must be adequately labelled. Any papers found should be handed over to the Police or the investigating team at the earliest opportunity. Documents that may be found include log books, passenger and freight manifests, maps, navigation logs, certificates of air worthiness, certificates of maintenance, technical logs, and many more.

Any mail sacks and pouches that are found should be protected from any further damage by removal to a safe and secure point such as the incident control and this information passed to the Police.

(b) Flight Recorders

All civil passenger aircraft as well as most others carry flight recorders known in the media as the "black box". They are indeed not coloured black, but invariably a brilliant fluorescent red (see Figure 1.17). The flight recorders are built to withstand severe shock and fire and may contain vital evidence as to the cause of the crash. It is therefore imperative that if they are found, they should not be handled unless absolutely necessary and preferably left where found. The Police or the investigation team should be informed of their discovery and location. They should only be moved if in real danger of being irretrievably lost, i.e., into a swamp. If they have to be handled, it should be done with extreme care and the exact position where they were found recorded accurately.

The preservation of evidence combined with accurate recording are fundamental in sound post accident management.

(c) Statement of Witnesses

The Police and the investigation team will have to gather all the statements from eye witnesses, survivors etc., in order to present them eventually to an inquiry or an inquest. Rescue and firefighting personnel who attended the incident will also be asked to submit statements and these should preferable be done when crews return to the fire station or prior to them going off duty. The statements should be taken or reports made out by individual crew members and should be done as soon as possible so that accuracy is maintained whilst memories are still fresh. It may take good leadership and management skills to gather statements from rescue and firefighting personnel after they have been working for what may have been a long time in harrowing and arduous conditions. However, the importance of their statements in piecing together the "jigsaw" of the accident will be vital to the investigation.

6.8.7 Decontamination of Personnel and Equipment

Aviation fuel, hydraulic fluids and certain oils may cause irritation or dermatitis if they come into contact with the skin. All personnel who have had fuels spilled onto them should decontaminate themselves with a thorough wash with soap and water as soon as possible and their clothing should be sent away for decontamination.

Personnel who have been handling bodies should wash down with disinfectant including their uniform and protective clothing, prior to it being sent for decontamination. Any gear that has come into contact with bodies, e.g., salvage sheets, body bags, body sheets, rescue and extrication equipment, etc., should also be disinfected and cleaned.

6.8.8 Ignition Sources and Evacuation of the Area

(a) Ignition Sources

The accident site should be a "No-Smoking" and "No Naked Light" area to eliminate or reduce the possibility of ignition taking place.

Other sources of ignition should be looked for and eliminated e.g., radios, generators, appliance engines etc., where there is a possibility that they may cause an ignition hazard. It must be remembered that fuel and vapours may travel some distance and gather downwind and downhill of the accident site.

(b) Evacuation of the Area

Any occupants of buildings and any occupants of vehicles in the immediate area of the accident should be evacuated by the safest route to a safe area where they should remain until the accident site and danger area is considered safe. All doors and windows in buildings surrounding the area should be shut with gas and electricity turned off and any domestic fires and naked flames extinguished. Evacuees may have to be accommodated for a lengthy period of time and logistical arrangements will have to be made to cater for them.

6.8.9 Post Accident Counselling for Rescue and Firefighting Personnel

All rescue and firefighting personnel who have attended aircraft accidents where there are a large number of casualties and fatalities will be affected by the incident in one way or another. In some cases, personnel may suffer post traumatic stress

disorder, which will affect them in many different ways. It is important to recognise this type of trauma and provide for it with help and professional psychological counselling.

All personnel attending a major aircraft accident should be debriefed with help being offered in overcoming any trauma or stress that may result.

6.8.10 Hazards and Personal Protective Equipment

Additional Hazards

An aircraft accident site is a potentially hazardous environment and it is impossible to detail every type of hazard that would be encountered, the following are indications of the hazards that may be encountered.

Acids (batteries), aviation fuels, hydraulic fluids, (skydrol), Mercury (from temperature bulbs), lead, flares, Fluoromasters (viton – used for seals), and many more. These present a particular hazard if they come into contact with the skin.

Man Made Mineral Fibre (M.M.M.F.) is a collective term used to describe many composite materials. These materials are commonly used in aircraft construction and present a respiratory and irritation hazard – especially if they have been subjected to heat or flame. This subject is dealt with in Chapter 8 – Special Hazards in Aircraft Incidents.

Personal Protective Equipment

Depending on the nature of the incident, Chemical Protective Clothing and breathing apparatus may be required for fire service personnel (see chapter 8, Section 8.4).

The security of the accident site will be the responsibility of the Police authority and they will restrict access to all unauthorised personnel.

All personnel who enter the accident site should be informed of the potential dangers and hazards. They should be provided with the necessary personal protective equipment which should include, goggles, surgical gloves, overalls (disposable type), adequate footwear (preferably wellington boots), and some form of respiratory protection.

Aircraft Incidents

Chapter 7 – Rescue Techniques

7.1 General

In any aircraft accident/incident the evacuation and/or rescue of the passengers and crew is paramount with the emphasis on speed. Firefighting, if it is necessary, must be carried out with safeguarding of the evacuation and the protection of the fuselage as its main objectives. The officer-in-charge must try to position his appliances with this in mind (see Chapter 5 Section 3). The method of rescue will depend on the type of aircraft, its attitude, e.g., wheels-up, nose-down, intact or breached, and where the fire, if any, has broken out. As mentioned in Chapter 6, the likelihood of passengers or crew lying injured outside the aircraft, and possibly some distance away from it, must not be overlooked.

7.2 Rescue Techniques

7.2.1 Gaining entry to an aircraft

(a) General considerations

In particularly serious crashes, an aircraft may hit the ground with such impact that it breaks up completely, and the question of forcible entry will therefore not arise, although pieces of wreckage may need to be cut through in order to release trapped occupants and facilitate ongoing rescue operations. In less severe incidents, however, an aircraft may remain substantially in one piece, or may break into sections which nevertheless retain their integrity. In such cases it will be necessary for rescuers to gain entry to the aircraft through one or more of the access points provided.

Where an aircraft is still in one piece, much will depend on whether evacuation is being carried out at the time of the brigade's arrival, and if so the extent of such operations. The fact that an evacua-

tion is going on must not prevent firefighters from entering the aircraft for rescue purposes, and to protect the integrity of the cabin interior, provided that they can gain access without hampering the escaping occupants. The evacuation of an aircraft will of course be primarily via the main evacuation slides. Some slides may not have deployed correctly however, or failed to operate owing to impact damage. Others may not be used by the occupants owing to an obstruction within the cabin interior or the outbreak of fire. These unused exit/entry points together with emergency hatches may prove useful to rescue and firefighting personnel in gaining access to the cabin interior without disrupting any ongoing evacuation, given caution is observed where slides have malfunctioned or failed to deploy correctly. (Chapter 1, Section 7.)

Aircrew are highly trained in emergency evacuation procedures and will endeavour to ensure passengers vacate the aircraft speedily in the event of an emergency. Cabin staff are normally responsible for certain sections of the aircraft and for those passengers contained within them. The "procedural" evacuation of an aircraft will however rely to a great extent on the integrity of the aircraft following an accident/incident.

Occasionally, depending on how hard the aircraft has landed, one or more engines may still be running, mainly to maintain power to depressurise, open doors etc. (This will also be the case in many ground incidents – see Chapter 5, Section 5.4.3.) Firefighters should beware of this possibility, and the dangers associated with aircraft engines (Chapter 5, Section 5.3.1). If they need to pass in front of or behind an operating jet engine, they should always keep at least 7.5m away from the intake to avoid being ingested, and at least 45m away from the exhaust outlet to avoid being burned (Figure 5.3). Special care should be taken at night.

The particular hazards associated with helicopters, and details of access points on such aircraft, are described in Chapter 3.

(b)   Doors

Details of aircraft doors are given in Chapter 1 Section 1.7.1, and examples are shown in Figure 7.1.

They can all be opened from the outside, according to the instructions marked on them. These should be carefully read, as failure to appreciate the correct method of opening a door could lead to considerable delay in rescue operations and possible injury to the firefighter.

If it is necessary for a firefighter to pitch a ladder in order to reach a door, he should be careful to avoid the risk of being knocked off by the operation of the door, which will probably open quite quickly. Figure 7.2 gives details of the heights that may be involved.

If a door is jammed, it will usually be worthwhile for firefighters to spend time trying to force it whilst at the same time attempting to gain access elsewhere.

(c)   Emergency stairs

Some doors may be fitted with stairs, which in an emergency can be operated from either inside or outside. (See Chapter 1, Section 1.7.2.)

(d)   Escape slides or chutes

As pointed out in Chapter 1, Section 1.7.3, the operation of a door from outside will not normally result in the deployment of an escape slide. The possibility of this cannot however, be ruled out, and, if a slide does deploy, it is likely to do so

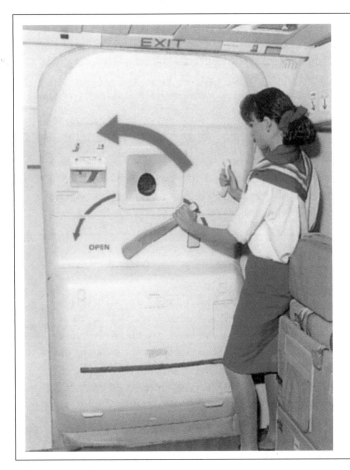 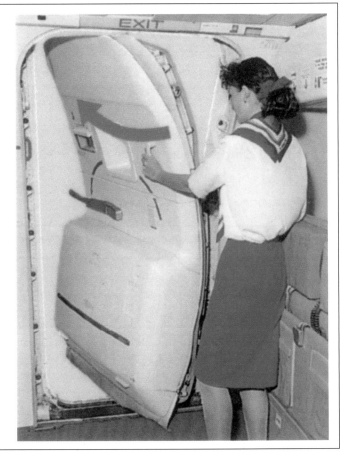

Figure 7.1 Operation of an aircraft door from the inside.

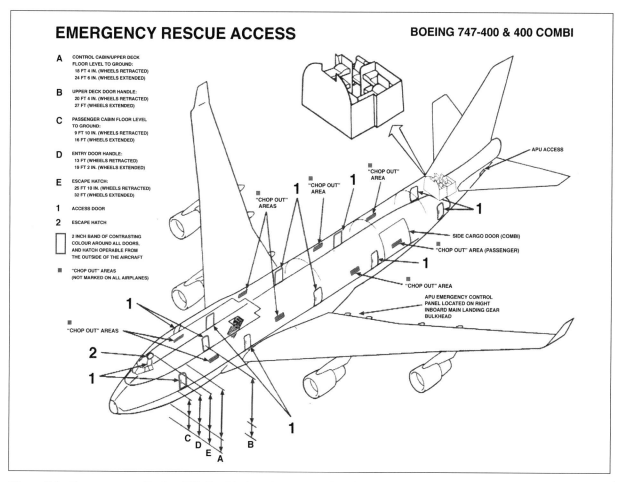

EMERGENCY RESCUE ACCESS — BOEING 747-400 & 400 COMBI

A CONTROL CABIN/UPPER DECK FLOOR LEVEL TO GROUND:
  18 FT 4 IN. (WHEELS RETRACTED)
  24 FT 6 IN. (WHEELS EXTENDED)

B UPPER DECK DOOR HANDLE:
  20 FT 4 IN. (WHEELS RETRACTED)
  27 FT (WHEELS EXTENDED)

C PASSENGER CABIN FLOOR LEVEL TO GROUND:
  9 FT 10 IN. (WHEELS RETRACTED)
  16 FT (WHEELS EXTENDED)

D ENTRY DOOR HANDLE:
  13 FT (WHEELS RETRACTED)
  19 FT 2 IN. (WHEELS EXTENDED)

E ESCAPE HATCH:
  25 FT 10 IN. (WHEELS RETRACTED)
  32 FT (WHEELS EXTENDED)

1 ACCESS DOOR

2 ESCAPE HATCH

▢ 2 INCH BAND OF CONTRASTING COLOUR AROUND ALL DOORS, AND HATCH OPERABLE FROM THE OUTSIDE OF THE AIRCRAFT

■ "CHOP OUT" AREAS (NOT MARKED ON ALL AIRPLANES)

APU ACCESS

"CHOP OUT" AREA

"CHOP OUT" AREA

"CHOP OUT" AREAS

SIDE CARGO DOOR (COMBI)

"CHOP OUT" AREA (PASSENGER)

"CHOP OUT" AREA

APU EMERGENCY CONTROL PANEL LOCATED ON RIGHT INBOARD MAIN LANDING GEAR BULKHEAD

"CHOP OUT" AREAS

Figure 7.2 Rescue access Boeing 747 Combi aircraft – various heights associated with access.

suddenly and with considerable force. Firefighters should bear this in mind, especially when on ladders.

If a malfunction has caused a slide to inflate whilst still in the cabin so that it blocks the door, firefighters should try to get it to deploy normally. If this cannot be done, it should be punctured so as to deflate it.

Escape slides should not be disturbed if they are being successfully used when rescuers arrive, but should be protected if possible and assistance given to people evacuating.

(e) Emergency hatches

Where passenger doors are impassable, the emergency hatches (see Chapter 1, Section 1.7.5) should be utilised to gain entry if they are not also being used in any evacuation. It must be remembered that these are heavy and require firm handling. Passengers using overwing exits may have to slide off the trailing edge of the wing or use the escape lines provided, and they should be assisted down to prevent injury.

(f) Breaking in

Cutting through the airframe (see Chapter 1, Section 1.7.6) should be a last resort. If this proves necessary, it will almost certainly entail the use of power tools. Firefighters must take great care that their use does not cause further injury to trapped or incapacitated passengers and crew. They should also remember that such tools, especially abrasive discs, can prove to be ignition sources in fuel-air atmospheres. When entry is made, any jagged edges must be suitably covered to avoid injury.

The underside of a fuselage is usually unsuitable as a break-in area because it only gives access to baggage compartments and/or cavities under the passenger floor which often contain cables, hydraulic pipelines, fuel pumps etc.

Entry into an aircraft is best achieved via a main door provided this in no way hinders any evacuation by passengers and crew.

Entry into the aircraft must be made as quickly as possible for rescue purposes and to safeguard the integrity of the cabin interior.

7.3 Operations inside the aircraft

7.3.1 General considerations

On entering an aircraft which has crashed, firefighters will probably be confronted with a very chaotic situation. Luggage will have cascaded from the racks, seats may have been torn loose and wreckage may be blocking the gangways. As stressed in Chapter 5, Section 5.3.2, any internal fire must be extinguished as soon as possible and the fuselage ventilated. Once control of any internal fire has been achieved, firefighting equipment should remain readily available to ensure any reignition, particularly during ventilation may be rapidly dealt with. Firefighters should not operate or interfere with any power controls or equipment unless absolutely necessary. The AAIB should be notified, via the police, of any alterations made to the settings of switches, levers etc.

It can be assumed that if passengers and crew have not evacuated they are either trapped, stunned, unconscious, injured, or dead. Firefighters must bear in mind the restricted room in the aircraft, and set up a system of releasing trapped survivors, removing them, if necessary along a 'chain' of rescuers, and taking them clear of the wreckage and out of any further danger. An orderly "in and out" arrangement should be adopted during rescue operations within the cabin interior given that sufficient entry/egress points are available. Such a system allows entry via one door and egress via another, facilitating the removal of casualties and the replenishment of equipment and personnel.

Civil aircrew are trained in rescue procedures and may be able to provide valuable assistance to firefighters in methods and routes of removal. In serious accidents, it will almost certainly be necessary to use power tools in order to reach and extricate trapped occupants, but firefighters should bear in mind the possible danger from these.

The procedures for casualty handling (see Chapter 6, Section 6.6.2) should be followed as far as possible. Persons suffering from asphyxia or haemorrhage will require urgent attention and should be removed without delay, or, if they are trapped, should be given immediate first aid where they are. It is important that firefighters provide and sustain a safe environment together with the tactical deployment of firefighting equipment within the cabin interior e.g., by foam coverage of any fuel spillage and ventilation of the fuselage, before the trapped or injured can be left in situ, following first aid, to await careful extraction and removal. The type of injury will to some extent dictate whether passengers are to be removed immediately or left until expert medical supervision is available.

Gloves should be worn by rescuers to avoid the danger of infection. Should personnel be scratched, immediate disinfecting treatment should be given.

7.3.2 Seats and seat belts

Passenger seats are designed to be easily adjustable, and crew seating on the flight deck also usually have a good measure of adjustment in their individual positions. Using adjustments of even a few millimetres could expedite the release of the trapped or incapacitated.

Seat belts for passengers vary in different types of aircraft, but they all incorporate a quick-release mechanism which is simple for the passenger to operate. They are unlikely to jam, but if for any reason this has happened the belt should be cut. Firefighters must be careful to cut away from the passenger, and this may prove easier if it is done at the side. Aircrew also have seat belts, which are usually different, generally incorporating a double harness, but are no more difficult to release.

7.3.3 Lighting

It may be necessary to provide lighting in the fuselage. Firefighters should be mindful of the fuel vapour danger and the need to sustain a safe environment. Trailing cables can become a danger and where they can be run in from a point not designated in any "in and out" arrangement, this should be accomplished.

It is important that firefighters provide and sustain a safe environment by the foam blanketing of any fuel spillage, ventilation of the fuselage and the tactical deployment of firefighting equipment within the cabin interior throughout rescue operations.

7.4 Treatment of Casualties after Rescue

Casualties (excluding the obviously dead) should normally be carried or led to an upwind (and if possible uphill) area designated by the Incident Commander, at least 50 metres from the aircraft if possible. This should be in a position accessible to ambulances or at least to medical personnel. The casualties should be protected from the elements with whatever material can be improvised. Cigarettes should not be offered to them in view of the risk that fuel may still be present in their clothing. Any necessary first aid should be given and as much detail as possible about injuries should be passed onto ambulance personnel before each casualty is taken to hospital.

If it is necessary to move any bodies away from the aircraft to prevent them from being destroyed, they should be kept well away from the designated casualty area.

7.5 Tactical Considerations

The Incident Commander at any major aircraft incident/accident will have a considerable number of operational and tactical measures to consider if passengers and crew are to be rescued and a successful outcome achieved. The following lists some primary factors to take into account:

Initially
- A dynamic risk assessment should be undertaken immediately on arrival.
- Consider the evacuation of all non-operational personnel to a distance of 300 metres.
- Ground slope and wind direction must be considered when siting appliances.
- Ensure Brigade Control are aware of exact location of the incident, the best method of approach and the operational situation.

Type of Aircraft

Military
- Smaller numbers of people involved.
- Dangers from weapons systems, armaments, explosives and nuclear radiation.

Civil
- Large numbers of passengers involved.
- Large fuel loads.
- Cargo unknown, consider hazardous materials, animal freight.

All Types of Aircraft
- Dangers from air intakes and exhausts of jet engines and the rotating blades of aircraft fitted with rotors and propellers.
- Caution to be exercised approaching any type of aircraft.

As Incident Develops

- Isolate fuselage from fire and heat.
- Attack from up wind.
- Spray foam along the line of the fuselage outward protecting the fuselage and escape routes.
- Secure the interior of the fuselage as soon as practicalbe, ensuring evacuation is not impeded.
- Full firefighting kit with breathing apparatus to be worn.
- Ensure safety of survivors, consider leaving seriously injured in situ until expert medical attention becomes available.

Chapter 8 – Special Hazards in Aircraft Incidents

8.1 General

Apart from the normal risks associated with fire-fighting and rescue operations following an aircraft accident, certain special hazards may arise. In the case of civil aircraft, dangerous substances may be encountered in freight; these include explosives, gases, flammable materials, and substances which are poisonous, infectious, corrosive, or radioactive. Military aircraft can present such hazards as bombs, missiles, small arms ammunition, and powerful laser target designators; there is even the possibility of nuclear weapons being involved. In dealing with any aircraft crash, firefighters should be aware of these possible dangers, in order to safeguard people and property in the vicinity of the incident and also to avoid exposing themselves to any unnecessary risks.

The suspected presence of dangerous substances or devices should not deter firefighters from searching for and rescuing survivors (whether inside or outside the aircraft) and conducting firefighting operations for the purpose of saving life. These remain their principal tasks and should be carried out with all possible speed in accordance with the advice given in Chapters 5–7. Firefighters should however proceed with caution, and in particular they should be careful to avoid touching or treading on any suspect equipment or material unless this is absolutely necessary, e.g., to extricate someone who is trapped. Any movement of dangerous items should obviously be done as carefully as possible.

Once all survivors have evacuated from the aircraft or been rescued, the incident commander will need to decide whether, in the circumstances, personnel should be withdrawn from the danger area immediately. It may in certain cases be preferable to let an aircraft burn rather than expose firefighters to further risk.

The presence of dangerous materials may sometimes necessitate the evacuation of nearby buildings.

8.2 Civil Aircraft

8.2.1 Identification of hazards

Under regulations devised by the International Civil Aviation Organisation (ICAO) and the International Air Transport Association (IATA), the captain of a civil aircraft must be given written information on dangerous goods which he is carrying, including any radioactive materials. In an emergency, this information would be conveyed to ATC, who would relay it to the emergency services. There is also a requirement on the operator of the aircraft to inform the country in which the accident has occurred of any dangerous goods being carried as soon as possible. Nevertheless, there may be some delay in obtaining this information, and firefighters may therefore need to rely to a large extent on any visible markings indicating the presence of dangerous substances.

To facilitate the identification of dangerous freight, the International Civil Aviation Organisation has introduced a regulation (ICAO's Technical Instructions for the Safe Transport of Dangerous Goods by Air) requiring the use of United Nations substance identification numbers, as carried on land transport vehicles. Further information about the UN numbering system is given in the Hazchem List 10 1999. Unlike land vehicles, however, aircraft themselves do not bear any hazard warnings or substance identification markings. The UN numbers are rather, marked on each dangerous container or package individually, together with an indication of the class of hazard. (There are special arrangements for crop-spraying aircraft; *Manual of Firemanship,* Book 12, Chapter 7 refers.)

The UN list is continually being updated, but developments taking place in the chemical industry may occasionally result in newly-developed dangerous substances travelling unmarked because they have not yet been added to the list. The absence of a UN number should not, therefore, necessarily be regarded as proof that a substance is safe.

All radioactive substances carry the Trefoil symbol.

In the case of UK-registered aircraft, the marking of dangerous freight is subject to additional regulations, and the UN number will usually be accompanied by labelling of a more detailed nature.

As explained, firefighters may receive notification from the air authorities that the aircraft is carrying dangerous goods; or they may notice UN numbers or other markings on packages or containers; or they may have some other grounds for suspecting that dangerous substances are present, e.g., they may find a freight manifest indicating this. If so, they should adopt the procedure appropriate to the type of hazard in question, as set out in the *Manual of Firemanship*, Book 12, Chapters 8–9, and Technical Bulletin 2/1993. Evacuation should be considered if there is firm evidence that highly dangerous material is present and that it is likely to endanger people in the vicinity – perhaps because of damage to containers resulting in the spillage or exposure of their contents.

If any dangerous substances are discovered or believed to be present but the full list of dangerous goods on board has not been received, the incident commander should immediately contact the airline concerned, via the police, and ask them to provide such a list as quickly as possible. In the meantime, firefighters will have to deal with the incident as best they can, relying on any markings found and any expert advice which they are able to call on.

8.2.2   Types of hazard

The classes of dangerous substances are listed in the Hazchem List 10 1999, paragraph 7. The ICAO regulations specifically forbid certain particularly hazardous items – for example, certain types of explosives – to be transported by air under any circumstances; there are nevertheless a wide variety of potentially dangerous materials which are not prohibited. The range of substances that might be encountered is so great that it is impossible to list them all individually, hence the need for caution.

(a)   Chemicals

The great majority of dangerous substances likely to be found are chemicals. The main categories of these, and the general procedure for dealing with chemical incidents, are set out in the *Manual of Firemanship*, Book 12, Chapter 8.

(b)   Radioactive substances

The safe transport of all radioactive material by civil aircraft is essentially governed by comprehensive regulations recommended by the International Atomic Energy Agency (IAEA). The regulations are directed to ensuring that safeguards appropriate to the nature and quantity of the material are incorporated into the design of the packaging. It is fully recognised that radioactive material could be involved in a severe aircraft accident, and the design requirements are accordingly devised so that the package should not constitute a hazard even in extreme circumstances. In the case of packages containing material with a high level of radioactivity, the requirements include tests which demonstrate resistance in severe impact and thermal environments and the provision of a fireproof trefoil symbol indicating the presence of radioactive material. Although a substantial number of packages of radioactive material are conveyed by air over the UK each year, only a very small proportion contain appreciable levels of radioactivity. Most are radioactive isotopes for medical and research purposes.

Despite the stringent safeguards, the possibility of an escape of radioactive material in an aircraft accident cannot be entirely ruled out. The properties and hazards of radioactive materials, and the procedures necessary when dealing with incidents involving them, are detailed in Technical Bulletin 2/1993. If there is any possibility that radioactive material has escaped, the officer-in-charge should ensure that the police invoke the NAIR scheme (see Glossary), especially if members of the public are likely to be affected. (The NAIR scheme does not apply to military aircraft).

(c)   Pressurised containers

Some liquids and gases, which are not in themselves dangerous, may be carried in pressurised containers dispersed throughout the cabin and aircraft holds may burst violently during a crash fire. Firefighters should therefore take the usual precautions for this type of hazard, including cooling the containers.

8.3 Military Aircraft

8.3.1   Identification of hazards

Many incidents involving military aircraft occur at airfields, and can normally be dealt with by the specially-trained firefighting personnel on the spot. (See Chapter 4, Section 4.6.1.) Where an aircraft crashes away from an airfield, however, the Local Authority Fire Brigade will be likely to be first in attendance. It is essential for firefighters with RAF, RN or USAF bases in their area to have some knowledge of military aircraft, and firefighters in other brigades should also endeavour to acquire some basic information about them. Should then, a crash occur in their area, they will be able to identify certain potential hazards and take suitable precautions during the initial firefighting and rescue operations. The general features of military aircraft are described in Chapter 2. The main types of weapons, and the precautions appropriate to them, are set out in following section.

In any incident involving a military aircraft, the military authorities concerned will be able to give, via the police, details of any particular hazards involved and any special precautions that will be necessary. However, depending on where the call originates from, this information may not be available immediately. If it is not, firefighters may need to rely to some extent on their own knowledge of military aircraft and equipment.

Military technicians will be sent to the scene of the incident as soon as possible in order to investigate it. They should be able to offer firefighters some on-the-spot assistance in identifying any hazards. (For the special procedure for nuclear incidents, see Section 8.3.2 (e).)

8.3.2   Types of hazard

(a)   Bombs

Bombs carried by military aircraft may be of the high explosive (HE) kind, but the majority are likely to be training bombs, which are less powerful, although they can still be very dangerous.

High explosive or training bombs do not normally explode on impact of a crashed aircraft if no fire is involved, since the fuses are not likely to have been set. Nevertheless, their behaviour is unpredictable, and there is always some risk attached to them. **Except for the purpose of saving or protecting lives, firefighters should not approach closer than 300m to an aircraft known, or suspected, to be carrying HE bombs.** Firefighters will be reasonably safe from direct blast effects at this distance, though they should still be ready to take cover for protection against flying debris.

If a crashed aircraft catches fire, any HE bombs may explode in the resulting heat, and firefighters should therefore proceed with the utmost caution if it is necessary to approach the aircraft.

The RAF uses training bombs of various types and weights. Some of these are filled with smoke-producing compound. The danger from an explosion of a bomb filled with smoke composition only is not unduly great, except where the explosion takes place close to people or property. The explosion of a bomb filled with flash composition, however, can be lethal. The intensity of light produced by some types of flash bomb can constitute a danger to the eyesight, and precautions should therefore be taken to protect the sight, even at a distance from the aircraft, if such bombs are known or believed to be present and the aircraft is on fire.

If, during rescue operations, any bombs are seen in such a position that they may become heated, they should be cooled with water spray. Foam, if available in sufficient quantities, should be used to fight any fire but should not be used for cooling bombs, because, although the water drained from the foam will assist cooling to a small degree, it will not significantly reduce the temperature. Should an RAF or USAF officer be present, his advice concerning the danger of explosion should be taken.

It is vital that, apart from cooling the bombs, no attempt should be made by firefighters to move or in any way interfere with them, whether the bombs have been subjected to heat or not.

(b)   Small arms ammunition

The risk of danger from small arms ammunition in an aircraft incident is remote unless fire occurs. If it does, fragments of cartridge cases and projectiles could be propelled up to a distance of 70 m. There is no risk of mass explosion, but small explosions are likely to occur with increasing frequency as the fire takes hold. These could be sufficient to cause wounds, especially where heavy calibre ammunition is involved, and it is also possible that the powder charge, or incendiary or tracer cores, may contribute to the spread of the fire.

The best protection is to keep low, avoid passing in front of the muzzles of guns, and have all belts of ammunition kept cool by water spray.

(c)   Rocket projectiles

Missiles are carried externally on pylons and these could break free from the aircraft during a crash. This would reduce the likelihood of the missiles becoming involved in any subsequent fire. If they do become heated, however, their warheads may explode similarly to HE bombs. It is less likely that the rocket propellant would be ignited, but if this were to happen the result would be unpredictable, because the missile might still be attached to its pylon, and the pylon might be distorted. It is unlikely, however, that the missile would be fired from the aircraft.

Missiles must always be regarded as potentially dangerous, whether a fire is involved or not, and it is important that any such weapons carried by a crashed aircraft should be located as soon as possible. Once this has been done, firefighters should try not to pass unnecessarily close to them, and in particular should avoid the area directly in front. In the event of fire, missiles should be cooled with water spray.

(d)   Pressurised containers

Virtually all military aircraft are fitted with pressurised containers of liquids or gases, with distribution systems for aircraft services (see Chapter 2, Section 2.4.2). Such containers may burst violently if affected by heat, but this will not occur in every case since it commonly happens that distribution piping is ruptured or melted and allows the pressure to be released before the contents of the container reach a critical temperature. Nevertheless, all pressurised containers should be cooled with water spray if the aircraft is involved in fire.

(e)   Nuclear weapons

The storage and movement of nuclear weapons and assemblies are governed by very stringent regulations. The inherent safety features, which form part of the design of nuclear weapons, ensure that there is no risk of a nuclear explosion should a weapon be involved in an accident. Nuclear weapons are not normally carried on combat aircraft in peacetime. They are, however, occasionally moved in special protective containers, secured within the hold of RAF and USAF transport aircraft. The risk of an accident happening is remote but, particularly during landing and take-off, there remains a possibility that one may occur.

In appearance a nuclear weapon is similar to a conventional high explosive (HE) bomb. All nuclear weapons contain HE when they are fully assembled and thus a conventional explosion may occur when a nuclear weapon is involved in an accident. Full precautions should therefore be taken as for an aircraft carrying HE bombs. An explosion, or any damage to the containers of a nuclear weapon or its components caused by fire, may lead to be spread, over a limited area, of radioactive dust and sensitised explosive (see Glossary).

In an incident involving an aircraft which is carrying a nuclear weapon, the RAF will send a Special Safety Team (SST) to the scene as soon as possible. The SST will be trained and equipped to deal with incidents involving nuclear weapons and assemblies, and they should arrive within a comparatively short space of time. It will be for the fire brigade however, to deal with the matter initially.

The RAF authorities notify all appropriate fire brigade controls, via the police, of any aircraft carrying nuclear weapons which they know to have crashed. The officer-in-charge of the PDA may therefore know at the outset, or receive early notification from his brigade control, that the aircraft is carrying nuclear weapons. However, this information may not reach him until after he has arrived and got to work. In view of the risks involved, officers-in-charge should proceed on the basis that any large transport aircraft of the RAF or USAF may be carrying a nuclear weapon unless they are informed to the contrary. If it is not known at an early stage whether a nuclear weapon is present, brigade control should report the matter to the police so that the RAF can be notified and asked for the necessary information.

Where a nuclear weapon is or may be involved, firefighters should keep well away from the aircraft and any debris, unless it is necessary for them to intervene in order to save or protect lives. The advice about the minimum safe distance for HE bombs (see Section 8.3.2 (a)) is equally applicable to nuclear weapons. If an approach to the aircraft is necessary, the appropriate procedures for a radiation incident (Technical Bulletin 2/1993) should be put into effect, including the wearing of breathing apparatus and the carrying of dosimeters and survey meters. The level of radiation should be constantly monitored until the arrival of the RAF SST. The approach should be made from upwind, and smoke should be avoided as far as possible since it may contain radioactive material.

If, in the event of a fire, a nuclear bomb or any component of such a bomb is found to have become heated, or seems likely to become so, it should be kept cool with water spray, like a conventional bomb. The spray should be kept in operation until the RAF SST Commander advises that it is no longer necessary. In the meantime, foam can be used in the normal way to tackle the remainder of the fire.

The RAF SST will be responsible for monitoring personnel for radiation, and for decontamination. When they arrive, the fire brigade incident commander should accept the advice of the SST officer regarding any further firefighting measures needed in the vicinity of the nuclear weapon and any necessary decontamination of personnel, uniforms and equipment. Firefighters or any other persons who have been within 300m of the crashed aircraft or debris, or who have passed through smoke issuing from them, should under no circumstances be allowed to leave the designated area until they and their clothing have been checked for radiation by the SST and, if necessary, decontaminated; nor should they eat, drink, or smoke until this has been done. The radiation hazard is alpha, which only constitutes a risk if contaminant is absorbed internally through cuts or by ingestion or inhalation. These precautions are designed to minimise this hazard. Because of the nature of the radiation involved, a check for radioactive contamination using the fire brigade's own radiac instruments cannot be considered sufficiently positive.

There may also be the likelihood of military weapons utilising depleted uranium both as a tip to weapons and as a balancing source on wing tips. Depleted uranium poses a similar hazard to rescue and firefighting personnel as any other radiation source.

8.3.3 Weapon Systems

Powerful laser target designators may also be fitted to some military aircraft. The biological hazard from lasers depend on the wavelength, the energy of the beam, its pulse length and the length of exposure. The eyes are the primary organs at risk from the laser's radiation but damage to the skin may occur if the energy and beam power density (irradiance) is high enough. There could also be a potential fire hazard with high power lasers. Other toxic risks may be present in relation to the medium or materials used in construction of the laser, e.g., some dyes, beryllium tube linings etc.

8.3.4 Military Aircraft Hazards Database

The Royal Air Force's Aeronautical Rescue Co-Ordination Centre (ARCC)'s database will provide information (24 hour availability) on hazardous material found on military aircraft. See Chapter 2, Section 2.10.

8.4 Man Made Mineral Fibres (MMMF)

As previously stated in Chapter 1, the use of man made mineral fibres, also known as composite materials, is common in the construction of aircraft both military and civil.

The use of these materials combine the strength and durability of woven fibres bonded together with resins forming a composite having 5 times the strength of metal; kg for kg.

The typical types of materials used are Carbon Fibre Reinforced Plastics (CFRP), Aramid Reinforced Plastics (ARP), Glass Fibre Reinforced Plastics (GFRP).

Toxicity

The risk to personnel arises from the decomposition of the material both during and after a fire. The intense heat usually found at an accident site will decompose the resins bonding the fibres liberating toxic isocyanate fumes. The fibres within the composite will break into shorter and smaller lengths increasing their respirability and transportability. There is the possibility that the material may plume following a crash and be carried considerable distances downwind. In addition to the respiratory risk, fibres can easily cause needle stick injuries and traumatic dermatitis.

Carbon fibres are capable of absorbing all the products of the post crash fire and if touched will act as an infection carrier enabling such products to enter the body. (See Figure 8.1.)

Conductivity

Airborne carbon fibre particles are conductive and if widely dispersed may damage electrical equipment and electrical installations.

Rescue

Owing to its inherent strength, rescuers may find it particularly difficult to cut through composite materials and heat generating cutting equipment has little or no effect.

Fire damage will however significantly reduce the structural strengths of composite materials and whilst floor panels may appear intact, they are unlikely to support the weight of a firefighter in such a condition.

Contamination

Three major areas of danger from composite materials found at crash sites have been identified. These include:

● Toxic vapours and dust released through the incineration of composite fibres.

Figure 8.1 Carbon fibre filaments and splinters remaining after aircraft fire.

- Sharp filaments or splinters of material distributed or exposed by impact.
- Gases released by burning resins.

In each category there is a significant danger from the ingestion or inhalation of vapour, dust or splinters from any wreckage.

The Incident Commander must ensure that safety measures are implemented which reduce the possibility of contamination not only to firefighters but to other emergency service personnel who may be involved in rescue operations within the crash zone. This must include the wearing of full firefighting kit with breathing apparatus during any firefighting operations.

The periodic application of foam will reduce the risk of airborne pollution by MMMF and the general disturbance of the material within the aircraft wreckage.

Personnel engaged in rescue operations should be kept to a minimum commensurate with the scale of operations.

Special decontamination procedures may need to be implemented for personnel who have been involved in any incident involving MMMF.

Safety

Full firefighting kit with breathing apparatus must be worn during firefighting operations. During subsequent clearing-up operations, chemical-protective clothing and breathing apparatus should be worn. Periodic application of foam reduces airborne contamination.

See Aide Mémoire, Appendix A.

8.5   Freight aircraft

One of the main hazards associated with freight aircraft is the cargo itself. It is reasonable to assume that in an accident situation the disturbance of freight can range from small items becoming unsecured to large containers or pallets releasing their contents (drums, cartons or packages).

One of the other hazards that firefighters should be aware of is that of livestock. Thoroughbred horses are a frequently transported by air as are other animals including wild game species being either imported or exported. The hazard that these animals present to rescue and firefighting personnel are obvious, especially if involved in an accident where either injured or crazed the animals may try to escape from their restrictive environment. The assistance of veterinary staff may be required in such circumstances. (See Figure 8.2.)

8.6   Aviation Fuels

Some of the risks of aviation fuels are described in Chapter 1, Section 1.4. These notes are additional information regarding safety of the specialised fuels such as Hydrazine (H-70).

Hydrazine (H-70)

In addition to the information provided in Chapter 2, Section 3, the following information is important for the safety of firefighters who may encounter this substance at incidents.

Toxicity

Inhalation of the vapours of hydrazine causes irritation to the throat, nose and respiratory system.

Prolonged inhalation of the vapours causes hoarseness, dizziness and nausea.

Contact with the skin can cause local damage or burning. If it penetrates the skin it has the same systemic effect as inhalation. Severe exposure can cause blindness. The Incident Commander must ensure liquid-tight, chemical-protective clothing with breathing apparatus is worn at aircraft incidents where the use of hydrazine is known or suspected.

8.7   Engines

As described in Chapter 1, section 3, the hazards associated with aircraft engines include those related to exhausts and intakes of jet engines and rotating propellors of turboprops and piston engined aircraft.

Figure 8.2 Thoroughbred horses being transported by air. The hazard that these or other animals present to firefighting personnel when an aircraft is involved in an incident is obvious.

There are however other less obvious engines that could represent a hazard to firefighters, especially during firefighting operations with engines or undercarriage situations. These engines are Ram Air Turbines (RAT's) or Air Driven Generators (ADG's) being designed to provide backup electrical and hydraulic power in the event that in-flight primary systems fail. (See Figure 8.6)

They are designed to deploy from flush mounted compartments in the belly of the fuselage or from similar flush engine storage compartments. They can be deployed with some considerable force being of substantial construction.

Fire service personnel should be aware of aircraft incorporating these devices and their respective locations.

Serious injury could result should these device accidentally deploy and operate.

Figures 8.3, 8.4, 8.5
Cargo aircraft may be
carrying a variety of
loads.
(Photographs courtesy of
HeavyLift Cargo Airlines Ltd.)

Figure 8.4

Figure 8.5

Figure 8.6 A Ram air turbine deployed on a Lockheed L1011. It is located in the centre underneath the fuselage slightly forward of the main landing gear.

Chapter 9 – Liaison and Training

9.1 General

The Civil Aviation Authority require all airports to promulgate emergency orders which amongst other things include the arrangements for summoning externally based emergency services. The Authority also recommends the establishment of an emergency services co-ordination group or liaison panel. The purpose of such a panel is to consider tactics, co-operative training, exercise planning, liaison and the promotion of greater understanding of the roles and responsibilities of each agency. The panel should include all emergency and supporting services liable to respond to or be influenced by an aircraft accident or incident. These include:

- Local Authority Fire Service and neighbouring Fire Service
- Airport Fire Service
- Police
- Ambulance Service
- Civil Aviation Authority
- Coroners Office
- Emergency Planning Officers
- Emergency Planning Working Groups
- Area Health Emergency Planning Departments
- Environment Agency
- Coastguard
- Airport Operations Management
- Airport Terminal Management
- Airline Companies

Local authority fire services should assign a liaison officer to the airport to form part of the liaison panel providing a consistent approach to emergency planning, training and operations. The role of the Fire Service Liaison Officer will vary greatly given the size and complexity of the airport to which ·he/she is assigned. The Liaison Officer should address the day to day matters at the airport such as operational training and communications as they arise. He/she should also consider the scope of risks posed by the airport other than those related directly with aircraft such as terminal buildings, hangars, fuel installations, underground or service rail transit systems (monorails etc.) and railway terminals. Additionally, areas surrounding the flight path immediately adjacent to the airport may pose operational or tactical difficulties, for example rivers, estuaries, or reservoirs.

The Liaison Panel should review emergency procedures regularly ensuring that not only are aircraft related accidents/incidents thoroughly examined but also the wider scope of risks addressed.

Local Authority Fire Services should assign a liaison officer to an airport. The Liaison Officer will sit on the liaison panel providing a consistent approach to emergency planning, training and operations.

9.2 Pre-Planning

The Airport Liaison Officer, together with other representatives on the liaison panel must ensure that each is fully aware of the arrangements for the mobilisation of all externally based emergency and supporting services to the airport in the event of an emergency. The plan must be agreed by all agencies and address fundamental areas, including:

- Size of aircraft and number of passengers
- The proximity of the airport, speed and weight of necessary response
- Water Supplies
- Access Routes
- Rendezvous points and Marshalling Areas
- Proximity of rivers, large areas of open water, reservoirs

Figure 9.1 Liaison panel meeting at Glasgow airport.

- County and National procedures
- Airport maps including operational areas, buildings, road traffic rules, Air Navigation Orders and regulations
- Communication
- Cargo, including hazardous materials
- Risk assessments
- Airport resources
- Command and control
- Pre-determined attendances
- Rescue and casualty handling
- Roles and functions of other services and agencies
- Compatibility of plans with other services and agencies
- Provision of escorting vehicles

9.3 Communication

Radios are essential for swift and accurate transmission of calls to respond to an emergency and for the control and co-ordination of any subsequent rescue and firefighting operations. At airports the Airport Fire Service will usually have radio communication facilities with both air traffic control and ground stations. The Airport Fire Service will also have a frequency to communicate with responding local authority fire services. At airports from Category 5 to Category 10, such facilities are mandated by the Civil Aviation Authority. However, at smaller airports radio communications are generally available to the airport fire service who will invariably have the ability to communicate with air traffic control. Airport fire appliances will usually have the facility to inter-communicate by radio and with their watchroom and/or control room.

It is essential that local authority fire services liable to respond to an airport are familiar with the radio procedures practised at airports as these are inherently different from those normally utilised by local authority services. In order to achieve good communications with the airport fire service responding local authority fire officers will inevitably need to manage two radio frequencies.

9.4 Training

Where fire brigades have a significant aviation risk, or where an airport falls within their response area, personnel likely to attend emergencies should receive appropriate training. This training should be commensurate with their respective roles and responsibilities.

The Liaison Officer assigned to an airport should co-ordinate an ongoing training programme in conjunction with the airport fire service.

This programme should ensure firefighters, watch/crew commanders, incident commanders, and specialist officers receive appropriate training.

The Liaison Officer should ensure that the training given, and the participation in emergency exercises, is recorded.

A number of areas should be incorporated including:

- Aircraft construction covering the types of aircraft regularly using the airport. This needs to be included both as basic classroom sessions and practical visits. A good knowledge of the terms used to describe parts of an aircraft, the numbering relating to engines and doors and so on will save a lot of misunderstanding when communicating with crew members.

- Special hazards posed by aircraft fires should be instilled coupled with the need for adequate personal protection.

- The tactics, techniques, and safe systems of work employed at an aircraft fire should be exercised.

- Joint training should be undertaken on a regular basis with the Airport Fire Service.

- The airport fire service, its role, responsibility and operational capacity.

- Pre-determined attendance procedures.

- Major incident procedures.

- The role of other emergency services and supporting organisations.

- Helicopter firefighting, tactics and techniques.

- Practical exercises and demonstrations.

- Floor plan exercises.

- Air Accident investigation branch, roles and responsibilities, operational procedures.

Communications can be a problem on an airport where radio transmissions by large numbers of personnel are usual. Valuable experience can be gained in carrying out communication exercises on the airport or in the immediate vicinity to diagnose potential problems and identify solutions.

Other measures and techniques which should be covered in training programmes are listed below:

- Aircraft entry and evacuation systems
- Water supplies
- Airport topography
- Aircraft familiarisation
- Post accident management

The subjects outlined are not intended to cover every aspect of aircraft rescue and firefighting procedures. They should however increase operational awareness of the hazards associated with aircraft and airport operations. Brigades will use a risk assessment to determine their particular training needs.

Training in conjunction with the airport fire service, particularly as it relates to practical evolutions will underpin inter-service co-operation and, indeed, may prove invaluable at the scene of an aircraft accident.

The Liaison Officer should establish an ongoing training programme for all personnel liable to attend an aircraft accident/incident.

Training, particularly practical training with the Airport Fire Service, may prove invaluable at the scene of an aircraft accident.

9.5 Exercises

An airport's emergency orders must detail the procedures for every emergency situation in which the airport fire service will be involved. These include, off airport attendances, domestic fires, chemical and radiological incidents, in-shore and off-shore rescue operations. The larger and more complex an airport the more comprehensive and elaborate its emergency orders will generally be.

In order to ensure the soundness of an airport's emergency plan it is essential it be put to the test. The Civil Aviation Authority require airports at Category 3 and above to conduct a full scale emergency exercise every 2 years. The exercise should involve the attendance of all externally based emergency and supporting services. The Liaison Panel should be actively involved in arranging and co-ordinating the exercise ensuring their respective agency's full participation. Where an airport is licensed to operate at night, alternate full scale exercises should be held during the hours of darkness. This requirement alone may pose considerable variations in tactical and operational methods which the Liaison panel will need to consider.

In intervening years a partial exercise should take place ensuring any deficiencies found during the full scale emergency exercise are reviewed and, where necessary, corrected. A table top exercise may prove appropriate in these circumstances with agencies using "real time" in mobilisation of resources.

Notwithstanding the requirements of the Civil Aviation Authority, the Liaison Officer should seek to ensure the plans of the local authority fire service are tested regularly. This may include small scale partial joint exercises with the Airport Fire Service possibly addressing an operational aspect such as water supplies, or breathing apparatus control. They may extend to table top exercises involving the primary emergency services focusing on mobilisation and communication. In developing closer working relationships at all levels, inter-service co-operation will improve, co-ordination between service groups will advance and communications strengthened. It is at the scene of a major aircraft accident/incident that the real benefits of good liaison will be realised.

In order to facilitate training of local authority fire service personnel, the Airport Liaison Officers Group have produced a booklet entitled "Initial Response to Aircraft Incidents". The booklet endeavours to assist local authority fire service personnel with fundamental operational knowledge in circumstances outside the immediate confines of an airport.

> The Liaison Officer should seek to develop closer working relationships, improve co-ordination between other service groups and strengthen communications.

9.6 Media

Whenever an incident/accident occurs at or in the vicinity of an airport, it will invariably attract considerable media attention. The Airport Liaison Officer should examine this prospect and have a rational strategy for dealing with it. This will normally be carried out in conjunction with the police.

Airports are generally very well equipped to manage media attention, being able to control access to operational areas through their strict security regimes. Many also employ public relations personnel versatile in handling communications with broadcasters and journalists. The utilisation of such resources may prove beneficial on occasions when emergency service resources are primarily focused on operation issues.

Chapter 10 – Safety

10.1 Managing Aircraft Incident/Accident Safety

Serious aircraft incidents/accidents are a relatively rare occurrence. Even fire services whose response area takes in a large airport have limited opportunities to build up experience of such incidents/accidents.

As a result, operational personnel are themselves unlikely to gain very much experience in aircraft firefighting.

It is therefore crucial that fire services have systems in place to ensure the safety of personnel who are to be committed to this infrequent and hazardous activity.

The key risk control measures that fire services will need to include to ensure firefighting safety must be pre-planned and include:

10.1.1 Risk Assessment

An assessment will need to be made that takes account of the likelihood and severity of any specific aircraft incident/accident. The severity will depend to a large extent upon location; town or remote countryside or at an airport. The size and scale of aircraft operations within a fire service's boundaries will in large measure influence probability.

The degree to which personnel will be subject to the risk of injury from incidents/accidents involving aircraft will depend upon the following factors:

- Crash rates
- Aircraft construction
- Fuel
- Other onboard hazards
- Manual handling

10.1.2 Liaison

Fire services will need to liase with a number of external agencies both nationally and locally. Agencies will differ depending upon locations, however, airport authorities/management should form a fundamental link with fire services whose ground encompasses their boundaries. In such circumstances a liaison officer should be assigned to the airport to form part of an emergency services co-ordination group or liaison panel. The officer will then be able to provide a consistent approach to emergency planning, training and operations. (See Chapter 9.)

10.1.3 Pre-determined Attendance and Resources

Having established the likely nature of any incident/accident or the type and size of aircraft involved, fire services will consider the level of response that would be appropriate. This will include the provision of specialist personnel, appliances and equipment.

10.1.4 Local airport procedures and collaboration

Significant aircraft incidents/accidents will require that all personnel and external organisations are fully aware of local airport arrangements for access, movement of appliances and personnel "airside", casualty evacuation, information, communication and so on. Collaboration with other fire services is likely to be required to ensure the availability of resources.

10.1.5 Specialist operational information

Fire services need to ensure as far as reasonably practical, that suitable and sufficient information relating to the hazards, risks and control measures is available to personnel at the time of an incident/accident. This may be achieved in a number of ways but the information must be clear, concise, current and relevant. The information may be of a generic or specific nature.

10.1.6 Training (see Chapter 9, Section 9.4)

Personnel who are likely to attend aircraft incidents/accidents either on or off an airport will require specialist training. This training will be based on the outcome of the risk assessment and, to an extent, on the information contained within this manual. Learning outcomes will be both technical and practical, being designed to satisfy the identified training needs of the individuals and crews involved.

To enable firefighters to gain experience in accessing and moving around in an aircraft fuselage and holds, real or simulated practical aircraft training should be carried out.

The information contained within this Manual provides firefighters with guidance relating to dealing with aircraft incidents/accidents. The information will also help fire services to preplan their organisational arrangements which will ensure, so far as is reasonably practical, the safety of operational personnel who have to deal with such unusual and potentially hazardous conditions.

10.1.7 Safe systems of work

(a) General

Safety procedures applied to normal operational incidents will apply to aircraft incidents/accidents but additional factors need to be considered when training and dealing with aircraft. These will include:

Entry into an aircraft's fuselage is usually via one of the main passenger doors. Personnel should exercise accessing and opening such doors using their equipment likely to be available e.g., ladders or hydraulic platform. The correct positioning of such equipment is essential in ensuring the evolution can be completed safely and without repositioning being required.

When approaching an aircraft consideration must be given to propellers which may be rotating and to jet engine intake and exhausts, as engines may be running. The movement and positioning of appliances and personnel in the vicinity of aircraft will give personnel confidence in their ability to tackle an incident.

Personnel should always wear hearing protection when working in the vicinity of operational aircraft. Officers should bear this in mind when controlling the deployment of personnel ensuring adequate means of communicating instructions are available.

Baggage and freight holds on aircraft can be very confined when loaded with containerised pallets specially shaped to the contours of the hold. When entering into an aircraft hold, firefighters must be aware of the mechanised rollers making up the floor and take care how they proceed.

Operations at an aircraft incident/accident will invariably require breathing apparatus or some other form of respiratory protection to be worn. Decontamination may also be called for. Exercises using breathing apparatus together with decontamination procedures during training sessions will provide sound experience for firefighters.

Whilst aircraft of identical models may appear very similar externally, their internal configuration can differ markedly. They may be fitted for passengers, cargo or a combination of both, which usually separates the main deck. Firefighters should make regular visits to aircraft to familiarise themselves with the varying arrangements and operating patterns.

Aircraft carry hazardous cargo which may include radioactive isotopes. These risks should be identified and procedures put in place to ensure firefighters receive adequate protection and information for handling such incidents.

An aircraft fuselage contains a large quantity of combustible materials in close proximity and in a relatively confined space. In a fire this can cause rapid flame spread and the generation of flammable gases very quickly. **Explosive mixtures may form within the fuselage with potentially hazardous consequences of backdraught or flashover given varying circumstances.** It is essential firefighters and incident commanders are constantly aware of these possibilities and tactics and techniques devised for personnel to be able to operate in such conditions.

(b) Airport Operations

All personnel must be aware of the potential hazards when "airside" on an airport. This is the operational area where aircraft taxi, park, defuel, refuel, land and take off.

The movement of all traffic within airside operational areas is controlled and monitored by Air Traffic Control (ATC).

All appliances and personnel must be escorted airside to and from the incident/accident site. Escorting vehicles will have direct radio communications with Air Traffic Control to facilitate taxiway and runway clearances etc.

This is applicable even though local authority fire appliances may be responding to an accident/incident on the airport located "airside". It is quite possible that although a major accident/incident has occurred at the airport, aircraft will continue to operate airside, landing and departing regardless of the accident/incident taking place. In such circumstances greater vigilance and caution will be called for.

It is vitally important that responding emergency services are aware of the potential hazards airside.

Brigades should ensure that a policy of appliances and personnel being escorted to and from areas within the restricted zone (operational airside areas) is in place.

(c) *Aircraft under inspection or maintenance*

Aircraft undergoing maintenance or refurbishment will usually be housed in an aircraft hangar. An aircraft on a short inspection or maintenance period may be installed in the hangar with a full fuel load.

The failure of a component during maintenance could result in the major release of this fuel in a relatively confined space. This apart, other volatile solvents, large amounts of electrical apparatus, gas cylinders, heaters together with maintenance platforms, all present potential hazards.

Aircraft hangars are therefore equipped with contemporary fire detection and suppression systems. These may include the discharge of high expansion foam capable of covering the entire floor area to a depth enveloping the aircraft.

Personnel should familiarise themselves with all such facilities at airports to which they may be required to respond.

Aide Mémoire

List of considerations for the Officer-in-Charge of a Post Aircraft Accident Scene

CONSIDERATIONS

- Will firefighters need to be committed to the post crash site?

- What protective clothing will be required by crews at the site?

- The atmosphere downwind of the aircraft may be highly irrespirable – remain upwind.

- Use of Breathing Apparatus.

- What areas of the aircraft have been damaged and are they likely to contain composite materials?

- Keep the number of personnel committed to the site to a minimum.

- Strict control of inner cordon.

- Thorough briefing of crews.

- Reduce risk of spread of composites:
  ◊ Fine water spray
  ◊ Water based suppressant
  ◊ Foam blanket
  ◊ Polythene sheeting

- Wet decontamination:
  ◊ Contain run-off

- Consult:
  ◊ Air Accidents Investigation Branch personnel
  ◊ Environmental Health Officer
  ◊ Specialist Cleaning Company

- Consider evacuation of nearby buildings

Glossary of Aeronautical Terms

Ailerons	Primary control surfaces at each wingtip which operate differentially (i.e., one goes up when the other goes down) to give lateral (rolling) control.
APU	Auxiliary Power Unit. An auxiliary generator usually powered by a small turbine and used to maintain batteries, services on the ground, start engines etc.
ATC	Air Traffic Control. A system used in the UK to regulate the movements of all aircraft, both civil and military.
Autopilot	A gyroscopically-controlled device which automatically maintains an aircraft on a predetermined heading and altitude, operating the aircraft's control surfaces.
Avgas	A type of gasoline fuel used in aircraft engines.
Avpin	Isopropyl nitrate. A type of fuel used in engine starter systems on some military aircraft.
Avtag	A type of kerosene fuel used in aircraft engines, more akin to gasoline fuel.
Avtur	A type of kerosene fuel used in aircraft engines.
Bogie	Type of undercarriage with four or more wheels to each leg.
Braking propeller	A propeller which can be reversed in pitch to create a braking effect, usually to shorten the landing run.
Canopy	The transparent fairing over a cockpit.
Centre section	The part of the mainplane which joins the fuselage.
Combi	An aircraft in which the proportion of passengers to cargo can easily be varied by removing or adding seats in order to achieve the most economical combination.
Ejection seat	A crew seat which can be fired from the aircraft, complete with occupant, in an emergency.

Elevator	The movable portion of the tailplane which provides longitudinal (dive and climb) control.
Fin	The fixed portion of the vertical tail surface.
Firewall	A fire-resisting bulkhead, usually between an engine and the remainder of the fuselage or wing.
Flap	A moving section on the trailing edge of a wing, which can be extended to improve performance of the aircraft for a particular manoeuvre.
Fuselage	The main structural body of an aircraft, carrying the mainplane, tail etc, and providing accommodation for the occupants and load.
I.C.A.O.	International Civil Aviation Organisation. Sets international standards and recommended practices for all aspects of international air transport published through a series of 'Annexes'.
Integral tank	A fuel tank formed by the basic structure, usually of a wing, by making a fuel-tight seal of spars, ribs and skin. Most common form of fuel tank on commercial aircraft.
Kerosene	Aviation fuel similar to paraffin.
LCC	Linear Cutting Cord. A canopy-shattering device similar to the MDC.
Leading edge	The front edge of a wing or aerofoil surface.
Longerons	Internally-place stringers running continuously along the length of the fuselage, to which other assemblies are attached, e.g., cabin flooring.
LOX	Liquid oxygen system found on military aircraft.
Mainplane	The major lifting surface of an aircraft.
MDC	Miniature Detonating Cord. An explosive cord incorporated into a canopy. When operated it shatters the canopy, usually prior to the ejection of the aircrew.
Monocoque	A common form of light aircraft construction in which an outer fabric skin, supported by light frames and stringers, is a primary load-carrying structure.
Nacelle	An enclosed structure usually containing an engine.
NAIR	National Arrangements for dealing with Incidents involving Radioactivity.

N.F.P.A. National Fire Protection Association. An independant organisation publishing codes and standards on fire and related issues.

Oleo An undercarriage leg in which the shock is absorbed by a piston moving up a cylinder containing hydraulic fluid or compressed air.

Overshoot
(1) A misjudged landing in which the aircraft touches down too far along the runway to pull up safely.

(2) The procedure of "going round again" if a safe landing cannot be made.

Payload That part of an aircraft's total weight from which revenue can be obtained, i.e., passengers or freight.

Pitch
(1) The angle of the propeller blades to the vertical swept area.
(2) Oscillation of the aircraft in rough conditions fore and aft.
(3) The distance measured longitudinally between corresponding points on aircraft seats.

Pressure differential The difference between air pressure inside an aircraft cabin and atmospheric pressure.

Pressurisation The process of making an aircraft interior airtight and maintaining the pressure inside it higher than that outside.

Pylon A streamlined fairing on a wing or a fuselage to carry a fuel tank, weapon etc.

Ribs Short supporting struts placed at right angles to the spars in a wing.

Rotor A system consisting of between two and five narrow wing-like blades carried radially on a single vertical shaft which, when rotated, produce lift, or, in the case of a tail rotor, stabilise the aircraft.

Rudder The movable portion of the vertical tail surface, providing directional control.

Sear A device to prevent the firing of an ejection mechanism.

Sensitised explosive The non-nuclear explosive used to detonate a nuclear weapon. In an accident it could be sensitive to shock and become radioactive.

Slab tail A tailplane which operates as a single entity to give longitudinal control instead of separate elevators.

Slat A small section of a wing leading edge which can be moved to improve airflow under certain conditions.

Slot	(1)	A gap through a wing leading edge, designed to improve airflow under certain conditions. It may be fixed, or may be formed by the operation of a slat.
	(2)	A time allocated to an aircraft through which it may depart from an aerodrome.

Spars The main support frames for a wing, running from the centre section to the wingtips or from wingtip to wingtip.

Sponsons Small attachments to the fuselage or wheels of an aircraft, often a helicopter, sometimes in the shape of a stub-wing, to accommodate wheel mechanisms, flotation apparatus etc.

Stressed skin The sheet metal covering of an aircraft, which is designed and stressed to take a load and to contribute to the total rigidity of the airframe.

Stringers Metal struts running horizontally along the length of the fuselage, spaced round the circumference of the main frames. There are also internally-placed stringers, called longerons.

S.T.O.L. Short Take Of and Landing; applied to aircraft with the ability to take off and land within a short distance, approximately one third of the distance required by conventional aircraft.

Tailplane The horizontal stabilising surface at the rear of an aircraft to which the elevators are attached.

Torsion-box The main load-bearing portion of a multi-spar wing, comprising front and rear spars, ribs and skin.

Vents Automatic valves which ensure a gradual equalisation of any difference between the inside and outside air pressure due to local weather conditions.

VTOL Vertical Take-Off and Landing. Applied to types of aircraft with the ability to reach an altitude of 15m within 15m distance of the take-off point.

Wing root The part of the wing that joins the fuselage.

Aircraft Incidents

Further Reading

Fire Service Guides to Health and Safety

Volume 1 – A Guide for Senior Officers
    ISBN 0 11 341218 5
Volume 2 – A Guide for Fire Service Managers
    ISBN 0 11 341220 7
Volume 3 – A Guide to Operational Risk
    Assessment
    ISBN 0 11 341218 5
Volume 4 – Dynamic Management of Risk
    at Operational Incidents
    ISBN 0 11 341221 5

Fire Service Manuals

Firefighting Foam
    ISBN 0 11 341186 3

Petrochemicals
    ISBN 0 11 341227 4

Communications
    ISBN 0 11 341185 5

HAZCHEM List 10 1999
    ISBN 0 11 341223 1

Initial Response to Incidents – Airport Liaison
    Officers Group

Acknowledgements

HM Fire Service Inspectorate is indebted to all who helped with the provision of information, expertise and validation to assist the production of this manual. In particular:

Mr Bill Savage, GIFireE, Aviation Rescue and Firefighting Consultant

Stn. O. W. Wilson, MIFireE, British Airports Authority Fire Service

BAA Plc

Divisional Officer Graeme Day, MIFireE

West Sussex Fire Brigade

London Fire Brigade

Chief and Assistant Chief Fire Officers Association

British Aerospace   (Mr T Mason)

GKN Westland Helicopters Ltd   (Mr A Harrison)

The Fire Service College

Royal Air Force, Headquarters Strike Command, Benson

Mark Edward Woodward, MIFireE

Carmichael International Ltd   (Mr M Day)

British Airways   (Mr J O'Sullivan)

Crash Rescue Equipment Service Inc., Texas   (Mr D Carmichael)

Ministry of Defence, Central Services Establishment

J. C. Ferrall, Aviation Consultancy Ltd